SCIENCE, ACTION, AND REALITY

EPISTEME

A SERIES IN THE FOUNDATIONAL,

METHODOLOGICAL, PHILOSOPHICAL, PSYCHOLOGICAL,

SOCIOLOGICAL, AND POLITICAL ASPECTS

OF THE SCIENCES, PURE AND APPLIED

Editor: MARIO BUNGE

Foundations and Philosophy of Science Unit, McGill University

Advisory Editorial Board:

VOLUME 12

RAIMO TUOMELA

Department of Philosophy, University of Helsinki

SCIENCE, ACTION, AND REALITY

D. REIDEL PUBLISHING COMPANY

A MEMBER OF THE KLUWER ACADEMIC PUBLISHERS GROUP

DORDRECHT / BOSTON / LANCASTER

Library of Congress Cataloging in Publication Data

Tuomela, Raimo.
    Science, action, and reality.

    (Episteme ; v. 12)
    Bibliography: p.
    Includes indexes.
    1. Science–Philosophy. I. Title. II. Series: Episteme
(D. Reidel); v. 12.
Q175.T783   1985   001'. 01   85-14305
ISBN 90-277-2098-3

Published by D. Reidel Publishing Company,
P.O. Box 17, 3300 AA Dordrecht, Holland.

Sold and distributed in the U.S.A. and Canada
by Kluwer Academic Publishers,
190 Old Derby Street, Hingham, MA 02043, U.S.A.

In all other countries, sold and distributed
by Kluwer Academic Publishers Group,
P.O. Box 322, 3300 AH Dordrecht, Holland.

Printed in The Netherlands

TABLE OF CONTENTS

Preface   vii

Chapter 1:   1
PHILOSOPHY AND TRANSCENDENTAL THINKING

Chapter 2:
THE MANIFEST IMAGE AND THE SCIENTIFIC IMAGE   10
I   Conceptualizing the World   10
II   The Stereoscopic View of the World   14

Chapter 3:
THE MYTH OF THE GIVEN WORLD, KNOWLEDGE, AND LANGUAGE   22
I   The Myth and its Constituents   22
II   What is Wrong with the Myth?   26

Chapter 4:
SCIENTIFIC REALISM - SCIENCE'S OWN PHILOSOPHY   37
I   Kant and Scientific Realism   37
II   General Arguments for Scientific Realism   40
     Appendix on Quantum Mechanics, Bell's Inequalities,
     and Scientific Realism   52

Chapter 5:
METHODOLOGICAL ARGUMENTS FOR SCIENTIFIC REALISM   65
I   The Theoretician's Dilemma and Scientific Realism   65
II   Theoretical Concepts within Inductive Systematization   77
III Quantificational Depth and the Methodological
     Usefulness of Theoretical Concepts   80
IV   A Scientific Realist's View of the Role of Theoretical
     Concepts   87

Chapter 6:
INTERNAL REALISM   95
I   Metaphysical and Internal Realism   95
II   Causal Internal Realism   106
III Picturing   115

Chapter 7:
  SCIENCE AS THE MEASURE OF WHAT THERE IS   124
  I   On the Various Kinds of Scientific Realism   124
  II Ontology and the Scope of the **scientia mensura**   129
      -thesis

Chapter 8:
  SOCIAL ACTION AND SYSTEMS THEORY   141
  I   The Conceptual Nature of Social Action   141
  II   We-intentions and Social Action   146
  III Joint Action and Systems Theory   151

Chapter 9:
  THE GROWTH OF SCIENTIFIC KNOWLEDGE   168
  I   Truth and Explanation in the Context of Scientific
       Growth   168
  II   A Pragmatic Account of Scientific Explanation   172
  III What is Best Explanation?   175
  IV   Inductive Logic, Epistemic Truth, and Best
       Explanation   189
  V   Scientific Realism and the Growth of Science   200

Chapter 10:
  SCIENCE, PRESCIENCE, AND PSEUDOSCIENCE   210
  I   The Method of Science   210
  II   Science and Prescience   215
  III Magic and Religion   220
  IV   Pseudoscience   226

Notes   236
Bibliography   257
Name Index   267
Subject Index   271

PREFACE

Were one to characterize the aims of this book ambitiously, it could be said to sketch the philosophical foundations or underpinnings of the scientific world view or, better, of the scientific conception of the world. In any case, it develops a comprehensive philosophical view, one which takes science seriously as the best method for getting to know the ontological aspects of the world. This view is a kind of scientific realism - causal internal realism, as it is dubbed in the book. This brand of realism is "tough" in matters of ontology but "soft" in matters of semantics and epistemology.

An ancestor of the book was published in Finnish under the title **Tiede, toiminta ja todellisuus** (Gaudeamus, 1983). That book is a shortish undergraduate-level monograph. However, as some research-level chapters have been added, the present book is perhaps best regarded as suited for more advanced readers.

I completed the book while my stay at the University of Wisconsin in Madison as a Visiting Professor under the Exchange Program between the Universities of Wisconsin and Helsinki. I gratefully acknowledge this support. I also wish to thank Juhani Saalo and Martti Kuokkanen for comments on the manuscript and for editorial help. Dr Matti Sintonen translated the Finnish ancestor of this book into English, to be used as a partial basis for this work. His translation was supported by a grant from **Suomalaisen kirjallisuuden edistämisvarat.** Finally, and as usual, I wish to thank Mrs. Auli Kaipainen for her help in preparing the manuscript typographically into something worthy of printing.

I wish to thank D. Reidel Publishing Company for the permission to include passages from my paper 'Truth and Best Explanation', **Erkenntnis, 22,** 1985, pp. 271 - 299. I would also like to thank publishing house Il Saggiatore for permission to use passages from my paper 'Science, Protoscience, and Pseudoscience' to appear in Pera, M. and Pitt, J. (eds.), **La razionalita della scienza.**

Helsinki, April 1985

Raimo Tuomela

CHAPTER 1

PHILOSOPHY AND TRANSCENDENTAL THINKING

1. Philosophy is sometimes characterized as the mother of the sciences. Having developed sufficiently the sciences parted company with their mother and started lives of their own. The analogy contains an element of truth. As a historical hypothesis it may be roughly true but cannot of course serve as anything like a defining characteristic of philosophy. But the idea of philosophy as the foundation of the sciences is also interesting from a conceptual point of view. It therefore merits some further comments.

Kant and many philosophers after him thought that philosophy could provide a solid epistemological foundation for the special sciences. This was to be due to certain general philosophical principles which are true for a priori reasons, viz. the principles can be known to be true independent of any appeal to experience. According to Kant (1787), what is known a priori is both necessary and strictly universal: "Necessity and strict universality are thus sure criteria of a priori knowledge, and are inseparable from one another" (B4). Thus Kant operates with a very strong notion of apriority. According to him, the proposition 'Every alteration has a cause' is an example of an a priori proposition (B3). (Note that as, according to Kant, alteration is a concept which can be derived only from experience, experience may play a small role even in the case of a priori knowledge.)[1]

Kant calls such a priori principles as are necessary to make experience possible **transcendental** (B25). More exactly, he calls "transcendental all knowledge which is occupied not so much with objects as with the mode of our knowledge of objects in so far as this mode of knowledge is to be possible a priori" (B25). However, knowledge about knowledge of objects, viz. "meta-knowledge", cannot be sharply separated from knowledge about objects, viz. "object-knowledge", not at least in Kant's philosophy (cf., e.g., Hintikka's, 1984, lucid discussion). To use an analogy, this is because the human mind viewed as a knowledge-instrument or measurement device always influences and modifies the objects it is supposed to "record" or "measure". Irrespective of this, Kantian transcendental knowledge is something a priori and necessary (apodictic). Thus such transcendental principles are a kind of immutable givens - they cannot be changed due to thinking or experience.[2]

Generally speaking, it is characteristic of a transcendental philosophy that some of its central philosophical principles are held to be transcendental in something like the above sense, which - according to our interpretation - can be attributed to Kant or at least to a brand of Kantianism. (Our main interest here is not so much to give a viable interpretation of Kant's philosophy as to try to discover a line of philosophical thinking which indisputably to a great extent derives from his philosophy.) The basic requirement of the existence of immutable transcendental principles is sometimes coupled with the view that there is something called the "nature" of human knowledge, which can be known only by some specifically philosophical and hence non-empirical, means (cf. Rorty, 1979a). This additional view of course fits well in with the idea that philosophy provides the foundations of all empirical knowledge. In addition to Kant, we may mention Peirce, Husserl, the early Wittgenstein, Russell, and Carnap as representatives of transcendental philosophy, conceived as above.

If a philosophical system includes transcendental principles, they give it a kind of firm, immutable foundation, we said. One of the main theses of this book is that there is no such foundation (cf. Dewey, the later Wittgenstein and especially Sellars). Philosophy cannot provide such a strong transcendental foundation of knowledge, nor can it therefore act as the mother of the special sciences in this strong sense. This of course does not prevent philosophy from standing in a close relationship with the special sciences and from cooperating with them. Neither does this exclude the existence of some kind of weaker, a posteriori transcendental knowledge or assumptions.

We may perhaps say that philosophy is a discipline whose aim is to study and to understand "how things in the broadest possible sense of the term hang together in the broadest possible sense of the term" (Sellars, 1963a, p. 1). The aim of the special sciences, on the contrary, is to understand in what specific ways things hang together. The distinction between philosophy and the special sciences lies in the dimension generality - specificity rather than in anything like the alleged a priori transcendental nature of philosophical theses and principles as opposed to the a posteriori character of the claims of the special sciences.

Considering the matter of transcendental principles from the point of view of man - a thinker - we can say that man is part of changing nature and culture. Human thinking - philosophical, scientific and everyday thinking - is mutable rather than based on some a priori fixed "Archimedean point" (see our comments on synthetic a priori knowledge in Chapter 6). In a way this is a lamentable fact, for were such an absolute transcendental foundation possible to have, it would make at least a philosopher's life much easier, for everybody would surely prefer to build on rock (transcendental principles). This remark in fact applies to any intellectual system and it also applies to man's emotions - think of magic and religion. (We shall indeed later in Chapter 10 consider magical and religious thinking from this viewpoint.)

Instead of adopting a strict transcendental point of view it is reasonable to think of people and their activities (including especially their thinking) in a plastic way. Thus we may view systems of thought in analogy with a ship being rebuilt, plank by plank, while at sea. While this anthropological viewpoint allows for foundational general principles (if you like, transcendental principles in a non-Kantian sense) those principles are mutable factual principles - and this remark about mutability applies even to principles of logic (cf. Chapter 3, Note 1)). Our main arguments against immutable transcendental principles will be given in Chapter 3 (but also see Chapters 6 and 10). There we will present the so-called Myth of the Given (in fact three kinds of givens) and argue against it. The Myth of the Given is in fact the assumption of the validity of certain transcendental principles. By refuting the Myth of the Given and by adopting a plastic anthropological view of man we can essentially block transcendental philosophizing in the sketched rigid sense. (The rejection of the three types of givens may perhaps not suffice to block all such transcendentalism but it will be argued to suffice for the purposes of this book.)

2. The rejection of immutable transcendental principles is - perhaps somewhat surprisingly - closely connected to the other

main theme of this book, viz. scientific realism and its defense. As we know, science characteristically postulates unobservable, transcendent entities and properties. But at least the
type of scientific realism to be defended below, viz. causal
internal scientific realism, does not take this fact to require
or even suggest the acceptance of transcendental principles (in
the indicated strict Kantian-like sense) nor the interpretation
of such postulated entities as unembodied spirits or anything of
the kind.

Briefly put, we shall advocate and defend - at least to
some extent - the following "grand" theses in this book:

(I)   Empiricism is untenable.
(II)   Idealism is untenable.
(III)   Metaphysical scientific realism is untenable.
(IV)   Causal internal scientific realism is tenable.

These general theses will be given more refined formulations in
Chapter 6, but for our present purposes the above will suffice.
That the theses (I) - (IV) are true or acceptable will be seen
to depend centrally just on the rejection of immutable transcendental principles. For, as we shall understand empiricism,
idealism and metaphysical realism, these doctrines do incorporate rigid transcendental thinking. Let us consider the matter in
a preliminary way here.

It is appropriate to consider theses (I), (III), and (IV)
together. Let me thus start from idealism. The kind of idealism
I particularly have in mind is (subjective) German idealism, a
kind of post-Kantian transcendental philosophy. Such idealism
makes a strict distinction between the sciences of nature (die
Naturwissenschaften) and the human sciences (die Geisteswissenschaften). This distinction is taken to be not only epistemological and methodological (cf. the corresponding distinction
Erklären - Verstehen) but also ontological, something related to
matter versus mind. Accordingly the human sciences are thought
to require idealism. This thesis is grounded on a certain interpretation of the Kantian distinction between the empirical self
and the transcendental self (cf. Kant, 1781/1787, esp. A341 -
348/B399 - 406). According to this interpretation, the transcendental self, serving to constitute the phenomenal world, is
something which escapes the ontological sphere of the natural
sciences (even if this need not be so in the case of the empirical self).

Idealistic philosophers typically regard the transcendental self as some kind of a priori determined "substratum" or foundation of the ego such that no cognizing and thinking is possible without this unifying entity. This kind of a priori notion of the transcendental self is accepted by Husserl (1913), for instance. In his philosophy the phenomenology of the Lebenswelt gives the subject a transcendental, a priori determined role. Also (the early) Heidegger, Gadamer and Apel are transcendentalists in some related senses. (Habermas seems to avoid the charge of a priori transcendentalism because of his "linguistic turn"; cf. Habermas, 1979.)

Generally speaking, the hermeneuticists think of man as a being capable of changing himself "from the inside", viz. as a being capable of redescribing and revaluing himself in various ways. This idea has often been seen as requiring an idealistic ontology and, indeed, something like the notion of the transcendental self. Thus, for instance, many philosophers think of human agents in the Sartrean way as existing primarily "pour-soi" (for itself) rather than merely as "en-soi" (in itself) and take (implicitly or explicitly) this distinction to be ontologically significant.

But it is reasonable to claim that no such ontological distinction is involved in the various concepts (such as the Sartrean ones) related to the self. Man does not have any special deep ontological essence which is some kind of "pure spirit". Another, and I think correct, way to think of the Kantian empirical self - transcendental self distinction is to distinguish between man as an empirical self and man as a moral agent. The notion of empirical self can be regarded as having ontological content - indeed it seems ontologically naturalistic and even materialistic (see Chapter 6, Section II for discussion). On the contrary the notion of a moral being is not an ontological one and thus in a sense does not add anything substantive into the world (see, e.g., Sellars, 1963a, pp. 38 - 40, 1972, pp. 62 - 90, and Rorty, 1979b, Chapter VII as well as Chapter 7 below). This solution entails giving up the notion of transcendental self (in the strong sense). Thus we have indicated how to conquer the strongest fortress of (subjective) idealism. (Also see the arguments of Chapter 7, Section II, against both subjective and objective idealism.)

The above solution to the problem of the transcendental self basically goes back to the distinction between describing

and justifying. It is arguable that discourse about justification, and prescriptive discourse more generally, involves no
ontological commitments. To this prescriptive "dimension" belong
also questions related to values and norms. Our solution entails
that one can connect different and incompatible sets of values
and norms with one and the same ontology. Talk about different
values is pertinent even in the case of philosophy and thus
philosophical views of man. Traditional Kantian type of view
emphasizes epistemic values such as search for truth and knowledge, and these values will be emphasized also in this book.
They need not, however, be the only values relevant to a philosophical view of man as thinker. One may thus regard, e.g., the
total welfare of mankind as an overriding and even supreme value
and add that truth is necessary for that. To mention yet another
and more different intrinsic value, the hermeneuticists' and
German idealists' favorite Bildung ("edification", "cultiva-
tion"). Bildung need not perhaps be regarded as incompatible
with truth, however, for at least in many cases search for truth
is necessary for achieving Bildung.

3. There is no need here to deal further with the theses (I),
(III) and (IV), as they will be discussed relatively much in
chapters to come. I confine myself to pointing out, briefly,
their connection to transcendental principles, starting from
thesis (I).

     In its classical form (Locke, Hume) empiricism contains the
idea that knowledge about real matters of fact is based directly
on man's senses: the limits of sense organs are at the same time
limits of the factual world. Applied to philosophy of science
the doctrines of empiricism typically generate an instrumentalistic view of science according to which - to put it very
roughly - scientific theories are mere instruments for the pro-
duction of observational (empirical) knowledge. I shall later
(in Chapter 4) criticize instrumentalism.

     My criticism draws heavily on the fact that (empiricist)
instrumentalism incorporates transcendental assumptions which I
labelled by the generic term the Myth of the Given. For empiricism (in its classical form) presupposes that the world is
causally given to us through non-conceptual and self-verifying
mental states (cf. Sellars, 1963a, pp. 156 - 161). Empiricism

takes concepts to be given similarly, too. According to the
empiricist doctrine of concept formation, so-called concept
empiricism, general concepts are abstracted from particular
sensations brought about causally by the objects and properties
of the perceptual world. Nothing else (e.g. other concepts or
"Vorwissen") is needed. As I will try to demonstrate in Chapter
3, these versions of the Myth of the Given have at bottom to do
with unacceptable transcendental principles.

Scientific realism, as it is mentioned in theses (III) and
(IV), can be characterized as follows. Scientific realism - in
contradistinction to at least strict empiricism - allows the
existence of things and properties which "go beyond sense experience". Such unobservable things and properties include electrons, wave-functions, and, to mention other kinds of unobservable entities, subconscious wishes, long-term memories, etc.
Scientific theories typically contain terms which express these
- in everyday sense unobservable - things and properties and
with the help of which they manage to talk meaningfully (though
perhaps in an idealizing way) about their topics of concern
(which may be sui generis). What there really is will in principle, and ideally, be found out by science, viz. the best-
explaining and true or approximately true theories as well as
the singular descriptions based on them and on the use of the
scientific method. Scientia mensura - science (or the method of
science) is the final arbiter of what there is. (This does not,
however, imply scientism - cf. Chapter 7.)

It is nevertheless of some interest to note that we can
distinguish several types of scientific realism on the basis of
the nature of the transcendental and other background assumptions involved. I shall in Chapter 6 examine the division between metaphysical and internal realism. I shall show that they
are separated from each other by the Myth of the Given. For
metaphysical realism assumes that the world, language, and our
knowledge are in a sense transcendentally given, while internal
realism denies these transcendental assumptions. I shall give
grounds for the rejection of metaphysical realism and develop
and defend a new variety of realism, causal internal realism,
which emphasizes the role of social and other human action.

It would of course be naive to think that a short book such
as this could sink once and for all broad traditional philosophical systems. Partly for this reason I shall focus on some
paradigmatic general versions of such doctrines. My arguments

cannot therefore be very detailed and exhaustive. This is an all
too common difficulty in philosophical writings where a large
number of fundamental issues are dealt with in scarce space.

4. Before starting with systematic examinations we perhaps ought
to have a short preview or rather a summary of the contents of
this volume. Chapter 2 deals with common sense thinking and
compares it with scientific thinking (and with thinking in
science). The setting is provided by Wilfrid Sellars' analogous
distinction between the manifest image and the scientific image.
Chapter 2 brings into the open most of the fundamental questions
of this work, and also develops some amount of conceptual equipment to deal with them.

Chapter 3 introduces the Myth of the Given, or more precisely, its ontological, epistemic and linguistic versions, and
gives grounds for their rejection. This chapter also discusses
Sellarsian semantics and argues that its adoption helps to
avoid the Myth of the Given.

In Chapter 4 I will bring forth general arguments to support scientific realism - in contrast with instrumentalism in
particular - and I accordingly characterize realism, in a sense,
as science's own philosophy. Chapter 5 continues the presentation of arguments for scientific realism in terms of various
technical developments having to do with the indispensability
(and usefulness) of theoretical concepts in scientific theories.

Chapter 6 examines metaphysical and internal realism and
develops an argument for internal realism which accrues from the
rejection of the Myth of the Given. In this central chapter I
shall develop and defend the mentioned causal internal realism.

In Chapter 7 we shall have a closer look at the contents
and scope of the **scientia mensura** -thesis. I will discuss so-
called minimal, moderate, extreme, and Sellarsian kinds of realism, which all in a sense serve to explicate the **scientia mensura** -thesis. I shall also discuss what this thesis does not
cover (having in mind here, in particular, axiological and
meaning-theoretic issues). On the positive side I shall briefly
argue for a kind of emergent ontological materialism. Thus we
will end up by tentatively linking causal internal realism with
a kind of ontological materialism.

Chapters 8 and 9 of the volume deal with scientific progress and the growth of knowledge from a pragmatic - or, better,

action-theoretic - point of view. Chapter 8 presents briefly the broad outlines of a theory of social action I have developed elsewhere, and uses systems theoretic tools to connect social action with the examination of social processes. That chapter contains material which - while it could not easily be left out - is not all needed for the developments in the chapters to come. But the general views and conclusions put forth there are central also for the rest of the book. Chapter 9 presents an analysis of the notions of best explanation and epistemic truth and argues for a realist view of the growth of knowledge, in part on the basis of so-called social practical inference. According to this view a scientific community can, by forming "we-intentions" and by acting in accordance with them, end up in a process where there is growth of knowledge, i.e. in a process which increases the explanatory power of theories and which in a sense leads to increasingly truthlike theories.

Chapter 10 examines in more detail the nature of science and scientific method - after all, scientific realism largely relies on them. In this chapter I shall also deal with the transcendental nature of magical and religious thinking as against scientific thinking. I shall also briefly discuss pre-science and pseudoscience here and criticize the latter in this final chapter.

CHAPTER 2

THE MANIFEST IMAGE AND THE SCIENTIFIC IMAGE

I. Conceptualizing the World

1. We sometimes speak of everyday thinking as well as of the everyday world (and world-view). This involves conceiving and conceptualizing the world in the way we have been taught to do in our own culture. The objects of the world are often primarily perceptual objects for us, or objects for our use, and we describe them accordingly. For instance, a table is typically regarded as a four-legged object with a relatively hard surface, which can be used for familiar everyday purposes. In contrast to this, we can think of the table offered to us by microphysics. In that picture the table is a wobbling and vibrating cloud of microparticles.

The everyday world, conceptualized in the above way, contains trees, dogs, houses, red balls, rose-scents, thinking, feeling and acting persons, schools, societies, and so on. A philosopher could say that the everyday world is populated in the main by Aristotelean entities and their properties. Wilfrid Sellars has labelled the everyday view the manifest image, and contrasted this with what he calls the scientific image (cf. Sellars, 1963a). Some distinction of this sort is unavoidable, and many like proposals have been put forth in philosophy. We shall briefly examine Sellars' dichotomy, one which suits our purposes well. Sellars' notion of the manifest image is an ideal type. Its conceptual roots lie in what he calls the original image. The original image is a completely anthropomorphic view of the world: all concepts appropriate for describing persons are thought also to fit inanimate things. The manifest image emerges as a result of the gradual depersonification of the original image, whereby inanimate things eventually cease to be conceived as thinking and acting agents.

The manifest image can also be given a positive characterization as the conceptual system or framework within which man became aware of himself as a person, and whose concepts he

normally employs when he examines himself. The manifest image is
by no means stable - there is continuous although slow development both in man and in the study of man (even though we have no
difficulty in understanding e.g. Aristotle's **Nichomachean Ethics**
today). The original image is refined, transformed and made
more precise, both conceptually (or categorially) and empirically. Note that Sellars includes within the means for empirical transformation only "inductive-correlational" methods (such
as Mill's methods and, more generally, methods for data analysis) and excludes theory formation based on the postulation of
new entities. In this book we allow the postulational method to
be used also in the context of the manifest image and thus we
let in current psychology and sociology, for instance, in so far
as they use the common sense framework as their conceptual
fundamentum.

Postulational theory formation is nevertheless very central
in science, and Sellars takes it to be one methodological characteristic of the scientific image. As an example of such
theory formation we can mention the explanation of the macro-
features of gases (for instance the Boyle-Mariotte law) with the
help of the kinetic theory of gases. In addition to this methodological characteristic the scientific image can be described
ontologically. According to it, the world consists of those
entities and properties which true scientific theories postulate
(cf. Chapters 4, 5, 6, and 7). (Since present scientific theories are at best only approximately true, we have here only
music for the future.)

2. Before we start comparing the manifest image and the scientific image, a few critical remarks are in order. The concepts
of both these images are vague. First, we can ask if the manifest image contains the concept of a star observable only by
means of a powerful telescope. (The everyday concept of star
presumably has not been formed postulationally.) Secondly, it
is unclear what fundamental or constitutive principles it incorporates. Does it for instance include a dualistic view of the
relationship between the body and the mind? Can free will and
the laws of the natural sciences be reconciled? These questions
hardly receive unambiguous answers based on the alleged "fundamental principles of everyday thinking" or the like. Thirdly,
as indicated, for example the social sciences in fact apply the

postulational method of theory formation, although they operate, in other respects, within the manifest image (e.g. factor analysis is often used postulationally). This of course shows that the Sellarsian notion of the manifest image is an ideal type, one which does not strictly speaking represent any extant "social practice"; and on this point we depart from Sellars' use of the phrase 'manifest image'.

It is often thought that the scientific image is a unified picture created by true scientific theories and one which we somehow approach asymptotically. Sellars refers in this connection to "limit science" in Peirce's sense, but he does not himself assume that there ever will be a Peircean community which can uphold such science. Nor does he assume that the scientific image could be given a unified description (see Sellars, 1968, p. 142). We can hardly give any firm a priori guarantee that such a unified, coherent scientific image will emerge from the theories of the ultimate limit science if there such be - and still less that we actually will reach those theories. (In this book we shall in fact make no assumptions about limit science nor use that notion; in our discussion the notion of best explanation in the sense of Chapter 9 will carry much of the burden of Sellars' limit science.)

The (or some) notions of the manifest and the scientific images are nevertheless philosophically useful, as long as they are used in a way which is independent from the above problems. In this chapter we are especially interested in the possibility of reconciling the manifest image and the scientific image.

3. The manifest image has been man's starting point in his attempt to examine the world and to formulate a scientific image of it. (This can at least be said when we speak of everyday thinking and scientific thinking, whose idealized products the manifest image and the scientific image represent.) The manifest image can be taken to be both conceptually and methodologically primary with respect to the scientific image. By this I mean that scientific vocabulary cannot have come into being in abstracto. Rather, it is in many respects analogical with the terminology of the manifest image or relies on it conceptually in some other relevant sense. The same goes for the foundations of theory formation. Thus we can claim that, for instance, the "microterminology" of the kinetic theory of gases has at least

in part been postulated on the basis of the "macroterminology" of the phenomenal theory of gases. This is clearer still in the social sciences (and psychology) where the postulational method is even more closely tied to the terminology of the manifest image - if indeed we want to include some areas in the social sciences within the scientific image (cf. neuropsychology and see e.g. Bunge, 1980).

Is the scientific image in any respect primary, then? Indeed it is. For it can be claimed, as scientific realists do, that the scientific image has been obtained by means of a method which is more reliable - and better in other respects also - than the means for obtaining knowledge within the manifest image. This method is that of science (cf. Chapter 9). Now, as a product of the full-blown method of science the scientific image then is (in an aposteriori sense) epistemically privileged. It is at least as central - and for some philosophers this is the most important thing - that the scientific image is ontologically primary. (If, however, ontology is understood in a Kantian fashion, ultimately as an epistemic category, the two mentioned aspects of primacy ultimately merge.)

The motto of scientific realism is, in a way, this: Scientia mensura - science is the measure of everything that exists. This is specifically an ontological thesis. Science is thought to be the final arbiter as to what there is and what there isn't. Such a view presupposes, of course, that the terminology of the manifest image is in no way logically or transcendentally irreplaceable and in that sense primary. This in turn presupposes the rejection of the Myth of the Given (in the sense to be specified in Chapter 3). Accordingly, due to that we can at best have indirect and (in a broad sense) theory-relative knowledge of the world; and, as said, best-explaining science will be the criterion of what there is. We shall return to the scientia mensura -thesis more closely in Chapter 7.

To use an old pair of philosophical concepts we can now speak of two "conceptual orders", viz. of the order of conceiving and of the order of being (the latter can also be called the causal order). The order of conceiving includes concept formation and the conceptual issues which have to do with knowing. Matters which have to do with occurring in rerum natura, with causal influencing and real existence, in turn belong to the order of being. We can now examine issues of primacy with respect to each dimension.

A typical empiricist (such as Hume) and a natural language philosopher (e.g. Moore) take the manifest image to be primary with respect to both the order of conceiving and the order of being, A "strong" scientific realist (e.g. Feyerabend) on the other hand takes the scientific image to be primary for both aspects. A "soft" scientific realist (like Sellars) considers the manifest image to be primary in the order of conceiving but the scientific image primary in the order of being.

II. The Stereoscopic View of the World

1. What if the manifest and the scientific images get into conflict with each other, in some sense? A philosopher who regards one of the images as primary in both orders need not worry about this, as his philosophy essentially depends on a single image. But for a soft realist such a conflict poses a problem, for the manifest and the scientific images have to be reconciled with each other to form some kind of a "stereoscopic" view of the world. Let us look into the matter a bit closer.

What kinds of relationships can there be, in principle, between entities in the manifest and the scientific images? Apparently these relationships can be of several kinds. The following four are central possibilities. **First,** an entity in the manifest image can be identical with some entities in the scientific image (e.g., a volume of gas with a suitable bunch of molecules). **Secondly,** an entity in the manifest image can be a representation or an appearance in the human mind of an entity (or entities) in the scientific image. **Thirdly,** it can be thought that it is the entities in the manifest image that really exist. If it is nevertheless maintained that corresponding to these entities there are some entities in the scientific image, the latter are understood as fictions.

Fourthly, a slightly different option is to think that the entities of the manifest image really are **systems** of entities in the scientific image, without presupposing that they are identical with such entities, viz. elements of scientific systems. On the contrary, it is then assumed at least prima facie that the entities in the manifest image can be on an ontological level which differs from the ontological level of their parts, that is, some entities in the scientific image. In order not to leave this option mysterious we need to note what is meant by a

system here. Let us say that a structure $\sigma = \langle K,E,R \rangle$ is a system
with respect to a set of entities or states O if and only if K
and E are subsets of O, representing the system's components and
environment, and R a set of relationships which represent relationships between the entities or states of K and E. For purposes of generality O is in this definition allowed to consist
of either objects or their states. (Otherwise we shall speak of
systems by using concepts to be defined in Chapter 8.)

According to the systems-theoretical option the entities in
the manifest image, thought of as systems of entities in the
scientific image, can well have new or "emergent" properties (or
characteristics) with respect to the properties possessed by the
entities in the scientific image. This option in fact contains
the possibility of constructing (on the basis of a part-whole
relation) various kinds of emergent (though not perhaps irreducible) ontological levels. Thus we can speak of physical, chemical, biological, physiological, psychological and social levels;
this construction of levels does not in itself relate to the
manifest image - scientific image distinction. But it can be
appended to it, as we already saw.

Of the options presented (and we could list more) the
first, second and fourth are possible ones for a soft scientific
realist. The third possibility suits an empiricist, an instrumentalist, or a natural language philosopher. Speaking on a
quite general level, the entities of the scientific image, viz.
the entities postulated by best-explaining science, must, according to scientific realism, really exist. To this stand a
scientific realist can either add that also the entities of the
manifest image exist or may exist (if he, for instance, accepts
either option (1) or (4); cf. our **weak** realism in Chapter 4) or
that they do not really exist (option (2); cf. **strong** realism in
Chapter 4).[1]

The second possibility is Kantian (with entities in the
scientific image replacing noumena) and it is accepted, for instance, by Sellars. What is more, Sellars adds - or so I read
him - the requirement that for every property, or rather for
every attribute in the manifest image there is a corresponding
attribute in the scientific image such that the attribute and
the counterpart attribute are functionally similar or play analogous roles. Sellars criticizes the first option for not allowing us to take into account such qualitative features of the
manifest image as, for instance, homogeneous colors (colored

patches) and rose-scents. As Sellars sees it, these features are emergent with respect to the scientific image. However, the fourth, systems-theoretic possibility allows such emergence without postulating new types of basic entities. In fact, the systems-theoretic alternative also makes it possible that the entities of the scientific image somehow causally produce the properties of the entities in the manifest image, construed phenomenalistically (cf. the second option). Therefore, if a scientific realist accepts emergence, the systems-theoretic approach is a promising line for him.

2. In Chapter 4 we shall discuss various kinds of scientific realism with respect to the ontological constructions just presented, and the alternative (3) will be criticized especially in Chapters 3 and 4. We must still briefly examine some problems in the attempt to reconcile the two images into a "stereoscopic" synoptic picture of the world. It is not very central in this connection how such objects of the manifest image as trees, stones, cats, gases, houses, etc., and their various properties should be ontologically construed. The biggest problems arise in the case of persons and societies. Let us have a brief look at them.

When we try to fit man into the scientific image we have to take into account the "mind-body"-problem with its several central subproblems. These can be grouped in a rough way into the following conceptual categories:

(a)   sensations ("raw feels")
(b)   cognitive processes
(c)   personhood.

(In Chapter 7 we shall give these formulations a somewhat more inclusive formulation, but these will suffice here.) Let us think of a man who observes a red ball in front of him. This observation has a propositional content, say, "This is a red ball". Observing as a propositional state is a cognitive state. A man who has never seen a ball and who does not have the concept of a ball cannot be in such a cognitive state. On the other hand a blind man who does master the concept of ball can be in such a state. A blind man - unlike a normal perceiver - nevertheless lacks the non-cognitive sensation which distinguishes perceiving from the mere having of the thought: "This is a red ball". This sensation can be intensional (non-truth-functional), but it is not intentional or even cognitive (and thus conceptual).

I shall not here try to give further and more detailed arguments for the distinction between conceptual and non-conceptual mental states, although the distinction is important (see e.g. Sellars, 1963a, p. 127 ff.). We are now interested in the ontological side of the issue. It can be claimed, and has quite often been claimed in recent literature, that cognitive processes (e.g. thinking, believing, wanting, willing, perceiving) are in fact learnt functional processes which have no qualitative content conceptually associated with them. In other words, these processes are characterized according to the roles they have in a man's mental activities and (overt) doings (see e.g. Sellars, 1963a, 1980, Tuomela, 1977, Chapter 4). This role ("software") can be realized in different ways (by different kinds of hardware), and it is natural to think that neural processes constitute at least part of this realization.

The so-called analogy theory of thinking, which many philosophers have endorsed since Plato, and which Sellars in particular has developed and defended, is a functionalist theory. According to it, speech and thinking are analogical activities, and the analogy is based precisely on their conceptual (or semantical) role (cf. e.g. Sellars, 1968, and Tuomela, 1977, Chapter 3). Thus the sentence "This is a red ball" is assumed to have largely the same (if not identical) role in a person's speech system as his thought episode "This is a red ball" has in his thought system.

To put it somewhat more generally, a functionalist analogy theory of this sort starts with the conceptual assumption that overt reason-based (or, which amounts to much the same, "meaningful") action, especially speaking, is antecedently understood and relatively unproblematic. The system of action concepts serving as the conceptual fundamentum also includes social action concepts. Indeed, it can be said that the framework of social action - which of course is to some degree culturally and socially conditioned - forms the conceptual foundation for the analogy theory of cognitive mental processes. The analogy theory construes, in this fashion, cognitive processes as theoretical episodes of a sort, in effect as episodes which occur when a person does not speak (although there are no **conceptual**

hindrances to this). Richer and more refined versions of the
analogy theory allow all those mental states and processes
which, for instance, psychological theories, based on modern
views of information processing, have postulated.

The analogy theory construes the **intentionality** of cognitive processes (e.g. that a man can think about non-existent
states of affairs and among other things about the future) on
the basis of the intentionality (essentially: intentional referring) of the corresponding language use, and thus it frees the
notion of intentionality from ontological commitments to mystical mental acts. It also claims that the intentionality of
speech contains nothing that could not be accounted for by using
the semantical categories of language (e.g. the categories of
meaning, reference and truth) which in this connection are taken
as unproblematic. If, then, someone believes that there is in
front of him a red ball, the intentionality of this state or
episode of believing is also understood to be analogous with the
believing-out-loud that there is a red ball in front of him.

What was said above sketches functionalism and the analogy
theory on an idealized and general level. My intention has been
merely to outline a research programme and an approach, and not
to develop and defend it in detail nor even claim that it is
capable of giving a complete account of the nature of conceptual
mental states and episodes.[2] Later, especially in Chapter 3, we
shall return to the problems of linguistic behavior and meaning,
and then, I shall hope, the analogy theory will receive further
illumination. I have specifically wanted to present an alternative which allows a materialistic ontology, and one which is
compatible with present, and, we shall hope, future neurophysi-
ology (cf. Chapter 7).

3. How about non-cognitive sensations? They can also be characterized in accordance with functionalism - at least in part.
Take as an example the above case of a man who perceives the red
ball. We can say that the man has a sensation (a state or an
episode) which (in part) is caused by the presence of that red
ball, and which also (in part) causally explains the man's
cognitive perceptual state in that situation. Now it can be
claimed that associated with this sensation there are non-
functional qualitative features, in our example the homogeneous
redness of the ball (especially if the ball is a partially

transparent,   painted,   solid rubber ball). This **qualia**-problem
of sensations is a fairly complex one, and I shall not discuss
it here,   except for suggesting that **qualia** seem to require the
consideration of man's "hardware" (his biological constitution).
Even if such categorical, non-functional qualities were
needed,   they do not, I think, make the reconciliation of the
manifest and the scientific images substantially more difficult.
For we   can   think that mental states (including perceptual
states)   are from an ontological point of view emergent brain
states or properties: brains are systems which have relatively
emergent macrostates with respect to their anatomic and neural
microstructures. Sensations are (perhaps) best understood as
such macrostates, and the qualities of these sensations as
properties of these macrostates (cf. Bunge, 1980, and below
Chapter 7). In any case, when we defend scientific realism we do
not have to adopt a very detailed view of the nature of sensations, at least as long as our motto **scientia mensura** is not
violated.

4. We have now finished our brief examination of sensations and
cognitive processes. There still is the category of personhood
or agency. How is it to be conceived? Although man is on one
hand a being which realizes and seeks to satisfy his inner wants
and impulses, he does, on the other hand, take into account
various **rationality standards** in his activity: He is a person in
a strict, non-qualified sense. Thus he is bound to make contact
with different standards of conduct - e.g. moral responsibili-
ties and rights - which he can either obey or disobey. The
category of personhood, then, consists of notions such as wanting, hoping, intending, believing, knowing, loving, doing,
obeying a norm, and being responsible. Human beings (or rather
the full-fletched members of the biological species **homo sapiens**) are persons in the sense that the conceptual system of
personhood applies to them. (Of course this involves that the
human beings to some degree at least conform to the rationality
principles inherent in that conceptual system.)
It is important to see that the mentioned concepts are
social ones. They presuppose that persons belong to some society
or other. Thus we can claim that the "we-attitudes" of a man
towards his group hold a central place in the definition of
personhood. Especially the so-called **we-intentions** ("we shall do

X together" and **mutual beliefs** and background knowledge are
important (see the analyses to be presented in Chapter 8). Concepts which have to do with social action and, as can be
claimed, therefore also all central social macro-concepts (e.g.
organization, institution) are for the most part based on them
(see Tuomela, 1984).

The concept of personhood is then tied to certain rationality standards, and these standards are, at least in some loose
sense, prescriptive, i.e., they concern the "ought" rather than
the "is".   It comes as no surprise, then, that the conceptual
system of personhood can in no way strictly be reduced to the
descriptive conceptual system of the scientific image. Behind
this claim there is of course the justified view that prescriptive concepts cannot entirely be reduced away by using purely
descriptive concepts (cf. Chapter 8).

The concepts of personhood, which form part of the conceptual framework of the manifest image in containing prescriptive
ingredients, thus to some extent differ qualitatively from those
in the conceptual framework of the scientific image. But we have
good reasons for claiming that such prescriptive elements bring
about no ontological commitments. When we build a stereoscopic
picture of the world we accordingly do not really reconcile the
descriptive machinery of personhood with the conceptual framework of the scientific image but rather join its **prescriptive**
concepts into the conceptual equipment of the scientific image
(cf. Sellars, 1963a, pp. 38 - 40).

Finally we must stress one of our starting points in the
building of the stereoscopic picture. It is that there are no
(ontologically, epistemically or semantically) a priori privileged - and hence irreplaceable - conceptual schemes. For if the
conceptual framework of the manifest image were irreplaceable in
just one of these areas, the formation of the stereoscopic picture would be impossible. Note that the primacy of the manifest
image in the order of conceiving is a matter of history or,
should we say, bio-psycho-sociology. It does not mean privilege
in the above a priori sense - and an asymptotic transition into
the stereoscopic picture in any case leads to the in-principle-
abandonment of the manifest image (in the final ideal situation
which at least in practice no doubt is unattainable). Giving
some conceptual system an a priori privileged position is to
accept the so-called Myth of the Given. As our scientific realism requires - among other things - the rejection of that Myth,

we have reason to examine this philosophically central transcendental myth in more detail in the next chapter.

CHAPTER 3

THE MYTH OF THE GIVEN WORLD, KNOWLEDGE, AND LANGUAGE

I. The Myth and its Constituents

1. Ordinary thinking takes it as natural that the external world has a clear and solid structure which, as it were, reflects in our minds and (indirectly) in our language. This is at least what an empiricist would say. In fact, when we look at the matter on this general level, it can be interpreted even more broadly as a true account. However, this way of thinking is often tied to the so-called Myth of the Given (or Givenness) which we have reason to think - as the term 'Myth' suggests - is philosophically untenable. What is the Myth of the Given all about?

One version of the Myth of the Given is based on the transcendental thought that the world has, for a priori reasons, a stable categorial structure: the world consists ultimately of certain kinds of entities (e.g. of trees, cats, men, tables, or, perhaps, sensa), of the different properties of these entities, and of relations between these entities. We could also say that according to this version of the Myth the world consists of "given" states of affairs or facts, the elements of which are just those objects, properties and relations. The world is, as it were, a priori sliced into "ready-made" entities and types of entities. I shall call this view the ontological version of the Myth of the Given. It can be given the following brief formulation:

(MG_O) There is an ontologically given, categorially ready-made real world.

2. What is typically referred to as the Myth of the Given in literature (cf. Sellars, 1963a) is what we shall in this book call the epistemic version of the Myth. This version contains the thought that the categorial structure of the world imposes

itself "primordially" on man's mind in somewhat the way in which a seal imposes an image on melted wax. Thus if someone is, via perception, immediately aware of an object which in fact is a brown table, he necessarily, in normal circumstances, recognizes it as a brown table. It is then assumed that one feature of such direct awareness (whatever direct awareness precisely amounts to) is its necessary (and incorrigible) connection to the world. More generally, the present idea of knowledge about the "external" world or internal states, given in immediate awareness, is this (cf. Sellars, 1980, p. 16):

(GIA)   If someone is immediately aware of an item which in point of fact (i.e. from the standpoint of "best explanation") has categorial status A, then he will be aware of it as having categorial status A.

It is typically thought that this awareness of the A-type item has been caused (in some suitable sense) by that item (cf. perceiving a brown table). The world thus causally produces knowledge of itself as a world of a certain kind, and its having been so produced is an a priori necessary feature of that awareness (epistemic state). This in a sense entails the thought that the world can causally produce knowledge (epistemic states) without any conceptual contribution by the knowing subject. These epistemic states can be called self-verifying or self-authenticating. Thus, according to the Myth, man can conceive the world as being of a certain kind with the help of self-verifying states without having any concepts about what it is to be of that kind. We can speak metaphorically of a mind's eye which has direct, concept-free access to universals.

I shall call this transcendental assumption concerning the possibility of antecedently non-conceptual, direct categorial awareness the epistemic version of the Myth of the Given. Generalizing the thesis (GIA) we can express it as follows:

(MG$_e$)   Man can be engaged in non-conceptual but yet cognitive epistemic commerce with the world.

Let me, furthermore, note that (MG$_e$) presupposes that either (cognitive) knowledge is non-conceptual and therefore concept-free or that the world necessarily causally produces in knowers a fixed non-linguistic "conceptualization" in the sense of (GIA) and does it, more generally, without the help of a prior set of concepts and prior background knowledge on the part of the knowers.

It is worth emphasizing already here that the rejection of (MG_e) does not speak against all causal accounts of knowledge - and we shall in fact endorse a kind of causal theory of knowledge below (see Chapter 6). The account of picturing to be defended there is just meant to elucidate non-cognitive epistemic commerce with the world, in a sense contradicting (MG_e).

3. The Myth of the Given also has a linguistic variant. At least in its empiricist form it contains the idea that the (semantic) meaningfulness of all factual terms is based on causal interaction of a certain kind between the language user and the extralinguistic world. An essential feature of that interaction is, according to the Myth, that there is a necessary (or "logical" or "ostensive") connection between language and the world. This connection can concern reference, meaning and/or truth. The so-called semantic concepts (truth included) are construed as relationships between language and the world, and these relationships are deemed necessary and hence privileged.

For instance, the observational word 'ball' can according to empiricism be thought to obtain its meaning and reference in something like the following way: particular balls cause in us certain non-cognitive sensations which bear a certain abstracted likeness to the balls, i.e. they are in some relevant abstracted sense ball-like. The word 'ball' - or any other word with roughly the same meaning (use), e.g. 'pallo' in Finnish - is then assumed to bear a necessary semantic connection to those balls via the mentioned class of sensations produced by those balls. (If you find this talk about necessary and immutable connections between a language and the world obscure, it may help to put the matter metalinguistically and consider sentences like "A ball is to be named by 'ball' (or its synonyms) in English". When opposing the linguistic Myth of the Given, the idea is to deny that such metalinguistic sentences are necessarily or "logically" true in any sense. Thus some other metalinguistic postulates could logically or conceptually have been adopted.)

According to traditional empiricism, observation words thus carry their meanings on their sleeves, so to speak. This view is called **concept empiricism**. According to it empirical concepts are created by abstraction from their empirical instances, and this brings about an irreplaceable conceptual (or "logical") connection between (at least) the observation language and the world. In more general terms, we are concerned here with concept abstractionism in which the basis for abstraction can in principle include also non-empirical instances, e.g. "evidential givens" or "intuitions" which are related to some more or less Platonic universals.

The linguistic Myth of the Given thus contains the transcendental idea of a logically and a priori privileged status of some language. This also means that a privileged language cannot be replaced by, or translated into, a semantically and ontologically different language without changing the subject matter of the language or without making discourse unintelligible. In empiricism this a priori privileged language is the observation language (a part of a natural language). In such a language, for instance, balls are somehow necessary represented by 'ball', 'pallo', and so on depending on the specific natural language in question; more generally, objects and properties have as it were their logically 'right' names. (Analogously one might think that according to some kind of metaphysical realism balls are necessarily connected to the semantic expression 'group of elementary particles XYZ'.) We can now formulate the linguistic version of the Myth of the Given as follows:

(MG_1) There is an irreplaceable, a priori privileged language (or conceptual framework).

We must still make it clear that although the forms (MG_o), (MG_e) and (MG_1) are compatible, they need not all be accepted or rejected at the same time. Thus (MG_e) may be true while (MG_o) is not. The former does not imply the latter - nor does the converse hold. Similarly neither of (MG_e) and (MG_1) implies the other, for conceptual and epistemic activities can be analyzed in a way which is at least logically language-independent. This should suffice as an exposition of the Myth of the Given. (It has been analyzed in more detail by Cornman, 1972; cf. also Sellars, 1963a and 1980.)

II. What is Wrong with the Myth?

1. Why, then, are the versions (MG_o), (MG_e) and (MG_1) unacceptable? Justifying their rejection is not all that easy, and I can
in this connection only give a somewhat superficial outline of
the matter.

(MG_o) presupposes that the world is, as it were, ready-made
and made to suit or to be conceptualized by some absolute "natural conceptual framework". If the world were like that, there
would be no Quinean worry about the inscrutability of references
about the indeterminacy of translations, or about the equivalence of theories (cf. e.g. Quine, 1969). For it would then
come as natural to assume that it can be truthfully described
without commitment to this or that conceptual scheme. But such
an assumption quickly leads into trouble.

One reason for the rejection of (MG_o) is the following. The
best description of the world is founded upon the best-explaining theory of the world, assuming there to be one (albeit not
perhaps a unique one). For this theory says what there really is
in the world and how the constituents of this world really
relate to each other (cf. our discussion in Chapter 7 of the
scientia mensura -thesis, and see Sellars, 1963a, pp. 170 -
174). If this view of the connection between describing and
explaining is accepted, the argument is easy to carry on. For
explaining is tied to understanding, and understanding in its
turn to a conceptual scheme (and with it to e.g. research interests and viewpoints; cf. Tuomela, 1980b, Putnam, 1981, and
Chapters 6 and 9 below). The categorial structure of the world
and therefore also the notion of the world, in a sense depends,
then, on the used system of description (see Chapter 6, Section
II). And this system of description cannot be regarded as some
kind of absolute "conceptual scheme of nature". The very notion
of such an absolute conceptual scheme is empty.

Putnam has used a geometrical analogy against the assumption of a ready-made world (see Putnam, 1978, pp. 130 - 133).
Let our "world" be a straight line. A line can geometrically be
constructed in different ways. In one such construction the
starting point is assumed to be a point, in another one a seg-
ment of the line. The end result in both constructions is a
line, but they slice the world (= the line) in different ways.
Putting it even more strongly than Putnam, we have here two
equally accurate and truthlike but non-equivalent theories of

the world. The non-equivalence in question involves that sentences like "There are points" is true in one theory but fails to be true or even meaningful in the other one. However, these theories are mutually translatable into each other in a truth-preserving logical sense (cf. Pearce and Rantala, 1982b). Because of this, Putnam's present example is, after all, perhaps not very convincing. I believe that better examples can be obtained from more holistic descriptive systems such as from the comparison between the manifest image and the scientific image (cf. Chapters 2 and 4). For instance, the concepts in the framework of agency do not seem mappable in one-to-one fashion to the concepts of the scientific image, as recent discussion of the mind-body problem convincingly demonstrates. Although the resulting mental and physical descriptions may be regarded as comparable (in some general sense of comparability ultimately explicable in terms of people's language-use) they cannot be correlated structurally to give type - type correlations, nor are they commensurable as to their meanings.

Arguments like the above show that if the world is assumed to be ready-made and a priori given, such a concept of world become empty, for it cannot, after all, be described in a variety of different ways, as assumed. Thus we have reasons to claim that the structure of the world is not ready-made and that the notion of the world is in a sense partly dependent on our conceptual scheme (see also Chapter 6, Section II on this dependence - independence issue). Note that this in itself of course does not rule out the objective existence of a world which influences us causally.

2. We then seem to have reason to reject the thesis (MG_o) and to accept its negation not-(MG_o). How about the thesis (MG_e)? Sellars is a renowned opponent of this thesis (see Sellars, 1963a, especially pp. 157 - 170, 1967a, pp. 351 - 353, and Cornman 1972). As he puts it, in the case of such classical empiricists as Locke and Hume the epistemic version of the Myth of the Given hinges on a kind of "epistemic fallacy". They, as well as many other empiricists, apparently thought that epistemic matters can be analyzed by means of, and reduced to, non-epistemic ones. A contributing factor in this fallacy is the classical empiricist concept of **idea** ("thought"), as this concept is ambiguous and unclear with respect to the distinction

epistemic - non-epistemic (see Sellars, 1963a, p. 156 ff.). The
classical empiricists simply were not clear about when their
word 'idea' (or 'thought') meant something epistemic and conceptual (such as proper thoughts) and when something non-epistemic
(such as sensations).

   But this important distinction cannot be overlooked, for
"in characterizing an episode or a state as that of knowing, we
are not giving an empirical description of that episode; we are
spacing it in the logical space of reasons, of justifying and
being able to justify what one says" (Sellars, 1963a, p. 169).
It is equally impossible completely to reduce epistemic matters
into non-epistemic ones as it is to reduce "ought" to "is". This
does not exclude the possibility of "evidential" (and "non-
criterial") connections between epistemic discourse and non-
epistemic discourse (nor between normative and non-normative
discourse).

   It is instructive to take a quick glance at a fallacious,
empiricist attempt to analyze the epistemic status of observation reports (cf. Sellars, 1963a, pp. 166 - 168). According to
it observation reports are akin to analytic statements in that
correctly making them is a necessary as well as a sufficient
condition for their truth. This idea, perhaps reasonable in
itself, proves false if the correctness of such an observation
report is founded, in a strict sense, on obeying a linguistic
rule. If for instance the correct making of the observation
report "This is a ball" is understood to equal, roughly, the
obeying of the rules for the use of the words 'this', 'is', 'a'
and 'ball' in normal circumstances, we immediately get into
trouble or at least into dangerous waters. For even a parrot
could satisfy this. So the empiricist adds that the obeying of
such rules presupposes that the speaker is aware of or knows
that the object in question is a ball. More specifically, the
correctness of the verbal act of describing is founded on a non-
verbal state of awareness which is considered to be self-verifying (cf. the remarks above before the thesis (MG_e)). But this is
a solution which is both mysterious (What could such awareness-
episodes be like? Do we really have them?) and circular in that
this solution only shifts the problem to another level, so to
speak. In this way the empiricist in fact ends up endorsing the
epistemic variant of the Myth of the Given, which, on the basis
of the line of thought given above, can be regarded as unsatis-
factory.

An obviously better solution is this. An instance of the observation report "This is a ball", when made in the presence of a ball, expresses true observational knowledge correctly if and only if it manifests the speaker's tendency to produce, in normal circumstances, instances of the observation report "This is a ball" in situations in which he observes a ball. Now this analysis still is wanting in that it does not exclude parroting linguistic behavior. We accordingly have to require that the observer (the speaker) **knows** that the instance of uttering "This is a ball" is, in normal circumstances, a symptom of the presence of a ball. The speaker cannot have observational knowledge in a full-blooded, proper sense unless he already has other knowledge ("Vorwissen"). In a nutshell, a person does not come to have the concept of a thing (e.g. a ball) because he has observed such a thing. Rather, he can observe such things because he already has that concept. More precisely, for any item x and for any feature F, one can notice that x is F if and only if one possesses the concept of F. (The word 'concept' must be understood broadly so as to allow for e.g. prelinguistic concepts.)

All knowing is thus, for conceptual reasons, based on concepts and other knowledge which the knower already possesses. Therefore one cannot be in non-conceptual but cognitive epistemic interaction with the world. We cannot distinguish between what is "given" to us and what we have "postulated" (cf. the debate concerning the separation of observation and theoretical terms and the "theory-ladenness" of the former). We cannot, either, strictly separate between our knowledge concerning our knowledge of the world from our knowledge of the world. Ergo, the epistemic Myth of the Given is to be jettisoned. (See also the criticism of concept empiricism in the next subsection.)

3. I formulated the linguistic version (MG_1) as the claim that there is an a priori irreplaceable privileged language (or conceptual scheme). By this I mean a language which deals with the real world in the strong sense that there are, between it and the world, necessary or, rather, analytic (i.e. meaning-based) relationships. They are irreplaceable - a language would lose its intelligibility and semantic functions without them.

(MG_1) is, however, wrong. Why? (MG_1) instantiates the so-called magical theory of meaning, to use Putnam's phrase (Put-

nam, 1981). Such a theory assumes that the referents of words
and other expressions somehow belong to them intrinsically (or
"logically"). They can be manifested through mediating, meaning-
carrying but non-linguistic mental episodes, as in concept empiricism, or they can be tied to an assumption concerning non-
linguistic intentional acts which are - as Brentano claims - in
an unmediated relationship with non-linguistic universals (i.e.
with something general and instantiable). In a theory of language of this sort it is typically assumed that meanings are in
man's "head" or mind (and that meanings determine referents).
This conception of language is, however, untenable (see the
remarks below and e.g. the weighty arguments in Putnam, 1975 and
1981).

As opposed to such theories of linguistic meaning there are
the use theories of meaning according to which the social use of
language is of crucial importance for meaning and reference. Use
theories often also stress the conventional and **contingent** (non-
necessary) nature of language. Consequently a use theory of
language can generally, and in a natural way, be appended with
the negation of the linguistic version of the Myth of the Given
by assuming that the semantic language-world -relationships (if
the theory needs them at all) are contingent. (See below, subsection 4, for our an exposition of Sellars' use theory, which
rejects the Myth of the Given.)

Let it be noted in this connection that concept empiricism
in particular can be countered in a more direct way, for it
relies on the thesis (**MG$_1$**). The empiricists, recall, do not make
a clear distinction between epistemic and semantic matters. For
this reason alone the remarks directed earlier against the
thesis (**MG$_e$**) can easily be used to rebut also the thesis (**MG$_1$**).
For the notion of abstraction in concept empiricism is of epistemic nature, in fact based on the above thesis (**GIA**) about
immediate epistemic contact. If - as there are reasons to think
- there are no such self-verifying states of awareness included
in that epistemic process, concept empiricism turns out to be a
mistaken doctrine.

As yet another criticism against concept empiricism we may
note here that this doctrine does not regard non-logical infer-
ential relationships as central features of concepts. Thus while
it should be obvious that an item's being (all over) red implies
that it is not green, concept empiricism seems unable to account
for this.

When (MG_o), (MG_e) and (MG_1) thus have been found faulty, we must on the same grounds also reject the so-called dual level view of scientific language which is based on the Myth of the Given. For the dual language view presupposes a sharp ontological, epistemic, and semantic separation between observational and theoretical terms.

4. As pointed out above the linguistic version, as well as the other versions, of the Myth of the Given can best be criticized on the basis of an alternative, competing view. We just noted that (at least many of) the use theories can be formulated without recourse to the thesis (MG_1). This is particularly clear in Sellars' theory of meaning which analyzes concepts and meanings basically in terms of rules (much as Kant and Wittgenstein did; cf. Sellars, 1967a). Let us briefly discuss this theory, whose basic ideas, at least, will be tentatively accepted in this book.

According to this theory semantic discourse (using 'means', 'refers', 'true', etc.) is metalinguistic discourse. And such metalinguistic talk is, in that theory, not relational at all. This means the following. Let us examine the claims:

(a)   'Mauno Koivisto' refers to the present president of Finland;
(b)   'Bachelor' means unmarried man;
(c)   'Red' (in English) means red;
(d)   "Snow is white" is true if and only if snow is white.

The semantic claim (a) concerns a proper name, the claims (b) and (c) are about predicates (general terms), and the claim (d) is about truth. All of (a) - (d) are in Sellars' theory assumed to be strictly internal claims within the metalanguage in question. They do not contain ontological commitments. We must particularly note that 'refer', 'mean', and 'true' do not express two-place relations between an appropriate linguistic entity and the corresponding entity in the world. (We do not have, for instance, "Means('bachelor', unmarried man)", in which the second member of this relation is taken to represent something non-linguistic.)

Rather, semantic terms such as 'mean', 'refer' and 'true' **classify** expressions and thus give them their use-meaning: mean-

ings are classifications of language uses (in other words, linguistic roles), where the uses of language referred to may be
taken to concern either thinking or communicating. (On the other
hand the fact that a term refers or the fact that a descriptive
sentence is true implies in Sellars' semantics that their occurrences manifest some factual "psycho-socio-historical" relationships, mainly or centrally relationships between language and
the world; cf. below.) It is worth emphasizing that in Sellarsian semantics the basic building blocks are singular items of
linguistic behavior. No universals are thus involved, even if
these singular "languagings" are classified according to the
roles they play in language.

To facilitate later discussions I shall now present a somewhat longer exposition of Sellars' (or Sellars-type) semantics,
especially its philosophical aspects. Let us consider a meaning
claim (cf. (b)) which has the general form

(e)   'S' in language L means p.

'S' is a word or sentence in the language L and p an expression
whose meaning (or role) the makers of the semantic claim (e)
know (or which they can actively use). In a Sellarsian semantics
an analysis and a clarification of (e) is given by the claim

(f)   'S' in language L is a ·p·.

where ·p· is a predicate (general term) which applies to all
those expressions (in any language whatever) which have the same
semantical role as p has in our language. Therefore dot-quotation '· ·' represents a kind of "equivalence operation" for role
abstractions. In the claim (f) the role of the expression 'S' is
expressed precisely by the dot-quotation ·p·.[1]

Let us particularly emphasize here the fact that Sellarsian
semantics rejects the appealing thought of interpreting the concept of meaning in (e) as a relation between 'S' and something p
understood as non-linguistic. Meaning (the phrase 'means') is at
bottom a specialized form of the copula ('is') and the copula is
not a relation word. Hence meaning is not a relation. Other
relevant considerations are that this Sellarsian view makes it
easy 1) to give up abstractionism (and consequently e.g. concept
empiricism) - abstractionism is precisely the view which takes
meaning to be a relation between language and the world (or

Platonic universals) - and 2) to avoid the well-known regress argument by Bradley (according to which, if meaning is a relation, its meaningfulness requires a new relation, and so on).

The role of the expression 'S' can be represented, besides using the operation '· ·', by specifying the semantical rules for this expression in L. While the dot-quotation operation is said to **convey** the role of 'S' in L the semantical rules governing 'S' can be said to **spell out** this role. The word 'semantical' is in this context understood in an unusually wide technical sense, and we can in fact speak of three types of semantical rules:

(i)   world-language -rules (put in a simplified way, e.g., "A speaker is permitted (licensed) to say "This is a ball" when there is, in normal circumstances, a ball in front of him");

(ii)   language-language -rules (e.g., "A speaker is permitted to infer from the claim "This is a ball" to the claim "This is a round object""");

(iii)   language-world -rules (to idealize, e.g., "A speaker is, in normal circumstances, permitted to kick a ball in front of him if he accepts as true of himself the intention expression "I will kick that ball""").

In suitable epistemic conditions such semantic rules can also be ought-rules (i.e. instead of using the notion of permission they can be formulated by using the concept of ought). Two dot-quoted expressions can now be said to be identical if and only if they stand for roles (in the sense specified by the definition of the dot-quotation predicate) which occupy the same place in a system of roles determined by a specific set of rules (cf. (i) - (iii)).

A role-carrier - e.g. an adequate expression of ·p·, say p+ - is closely related to rule-following: an expression which contains p+ must be, when produced as a ·p·, in accordance with the semantical rules for ·p·. This rule-following can be based, in a strong sense, on a speaker's current intention to follow them. But there also is plenty of spontaneous rule-following linguistic behavior which in fact need not be even intentional action. Such behavior includes **pattern-governed behavior** which is based on habitual behavioral patterns. As an example we can take children's candid linguistic behavior, which is not always in-

tentional.   Such behavior is typically pattern-governed behavior
brought   about by the linguistic guidance provided by a   child's
parents (or linguistic peers).

But   we may also include more rudimentary and (probably) to
a great extent genetically based behavior here.   I have in   mind
especially behavior based on **sensory-motor schemes** in something
like Piaget's sense.   Such behavior (e.g. a baby's reaching for
its   mother's breast) is pre-linguistic, yet in a   broad   sense
conceptual and meaningful.   Pattern-governed behaviors of   this
type cannot be governed by ought-to-do -rules, but they may in a
sense obey ought-to-be -rules (cf. below).

We   can   illuminate the situation with an example which   is
connected to the meaning postulate (b). The analysis (f)   gives
us the claim:

(g)   The occurrences of the word 'bachelor' in English (and of
        'poikamies'   in   Finnish)   are occurrences to   which   the
        predicate ·unmarried man· correctly applies.

We   then tie (g) to the ought-to-be -rule between   language   and
the world which "governs" pattern-governed behavior. To simplify
a great deal, the rule is here (roughly) of the form:

(h)   It ought to be that the speakers of our linguistic commu-
        nity produce the expression (or that they tend to produce
        the   expression) "He is a bachelor" in the presence of an
        unmarried man,   under normal circumstances, when they are
        asked to describe the situation.[2]

The ought-to-be -claim (h) implies according to Sellars (but not
unproblematically; cf. the criticism in Tuomela, 1977, pp. 53 -
58) an ought-to-do -rule which is associated with   intentional
action and which is directed to the members of the   community,
especially to its language teachers:

(i)   The   members of the linguistic community ought to act   so
        as to bring it about that the speakers produce   the   ex-
        pression (or tend to produce the expression) "He is a
        bachelor" in the presence of an unmarried man,   in normal
        circumstances, when they are asked to describe the situ-
        ation.

We can say that the appropriate following of ought-to-do
-rules (cf. (i)) causally brings about, through appropriate
intentional action, such a result in the language-pupils of the
linguistic community that these pupils obey in their pattern-
governed behavior the corresponding ought-to-be -rule (cf. (h)).
The said result from learning is an appropriate disposition (and
ability) to produce linguistic behavior in the pupils. This
disposition can be a highly complex one, and we can presumably
think that it has some sort of a neural basis, which it is the
task of neurolinguistics to find out. The existence of this
disposition and its realizations in linguistic and other behavior form the primary ground for the thesis (g): without these
(g) cannot hold. (It is thus a kind of "criterion-relation", to
use Wittgenstein's terminology.)

Generally speaking, underlying Sellarsian semantical world-
language, language-language, and language-world rules there are
complex naturalistic uniformities of behavior: "Espousal of
principles is reflected in uniformities of performance" (Sellars, 1963a, p. 216). Thus language is connected to the world by
means of complicated naturalistic (and non-conceptual) psycho-
socio-historical connections which are at least to some degree
lawlike. Accordingly, we get naturalistic explanations of the
truth of statements made by people. For instance, the truth of
the statement-utterance 'This ball is red' by a person is explained, in normal circumstances, by this ball's being red (a
fact about reality). It is a complex matter to justify this explanation but in any case such justification would refer to the
mentioned psycho-socio-historical connections (relative to the
speaker) and it would spell out that the symbol 'a' represents
or pictures the object a in virtue of being concatenated to the
right with a token of the linguistic phrase 'is red'. (We shall
return to the central, naturalistic notion of picturing truth in
Chapter 6, Section III.)

As we have managed to show, Sellarsian semantics does not
rely on abstract objects or universals (such as concepts or
sets) in some Platonic or Hegelian sense (cf. the transition
from claim (e) to claim (f)). On one hand it is naturalistic,
for it relies heavily on linguistic behavior. On the other hand
it takes linguistic rules rather seriously. Although semantical
claims like (a) - (f) are not prescriptive nor descriptive in
the causalist sense of the word, they nevertheless rely on
prescriptions or rules via the notion of role.

5. I shall not in this connection justify or defend the Sellarsian theory or meaning.[3] I merely want to conclude by observing
its connection to the thesis (MG_1): Sellars naturally rejects it
and accepts its negation not-(MG_1). The crux of the matter is
that although in Sellars' system language of course is about the
real world, there are no semantic relations between language and
the world, for, we recall, semantic words like 'means', 'refers', etc. do not express relations at all. But on the other
hand, there are relevant "psycho-socio-linguistic", non-semantical and non-conceptual relations. These "Ersatz-semantical"
relations are causally produced via language-learning, which is
based on social practice. In other words, (natural) language
relates to the world through causal regularities in language
use. We have particular reasons, therefore, to emphasize that
these regularities are non-conceptual invariances in the causal
order in the first place. Thus although the proper name 'Mauno
Koivisto' does in a sense refer (viz. Ersatz-refer) to the
current president of Finland, this reference is based on causal
and non-conceptual interaction between the use of the proper
name and the person named.

One can in fact reject the thesis (MG_1) even if one thinks
that the terms 'refer', 'mean', 'true', etc., somehow represent
relations between language and the world. But one has to take
these relations to be in some sense contingent and replaceable.
Such a non-Sellarsian view can be made understandable and acceptable, especially if one clearly rejects the classical division of sentences into analytic and synthetic ones (i.e. if one
in fact rejects analytical relations in a strict conceptual
sense).

The end result of this chapter is, in nuce, that the Myth
of the Given, or in fact its versions or types (MG_o), (MG_e), and
(MG_1), can and should be rejected. At the same time we have
discovered good grounds to reject all philosophical doctrines
which rely on the Myth of the Given, or any of its discussed
versions. Classical empiricism clearly is such a doctrine, and
consequently we have reached one of the goals presented in Chapter 1 of this book. (Some versions of intuitionism and phenomenology also rely on the Myth of the Given in some of the senses
distinguished. Husserl's, 1913, phenomenology can be mentioned
as an example, although we cannot here give further documentation; cf. Chapter 1.)

CHAPTER 4

SCIENTIFIC REALISM - SCIENCE'S OWN PHILOSOPHY

I. Kant and Scientific Realism

1. In his work **The Critique of Pure Reason** (1787) Immanuel Kant
sketched a philosophical system which since then has greatly
influenced the formation of different kinds of philosophical
theories. Especially many idealistic views have in one way or
other been based on Kant's philosophy. What interests us here is
the fact that some versions or types of scientific realism can
be developed from some basic Kantian themes and ways of thinking, albeit with rather radical alterations.

As we know, Kant thought that our "everyday" world, i.e.
the world conceptualized as consisting of trees, cats, balls,
red and round objects, acting and thinking persons, schools,
etc., is in fact a phenomenal world. This manner of speech gives
us access to the distinction between phenomena and noumena.
Noumena are "things in themselves" and phenomena the things as
they appear to us. We can say that phenomena in a strict sense
are "experiential" things (e.g. observations or things "contained" in observations) produced by noumena, while phenomena in
a wider sense - in the sense intended by Kant - are things in
the phenomenal or empirical world in the sense indicated (to put
it in the Kantian way, things which exist "actually" but not "in
themselves"). We must emphasize that the world conceptualized
phenomenally (empirically) in this latter sense, is dependent on
human mind, viz. its things (and properties) are in a sense
conceptually mind-dependent. Nevertheless things conceptualized
as phenomena actually exist.

About the noumena "behind" the phenomena, i.e. about the
things in the world as they exist independent of us, we can
according to Kant have no clear and distinct concepts or conceptions. We can know nothing specific about the noumena but at
best achieve some very general knowledge about them. The phenomena are spatio-temporal and subject to the categories of our
thought and to the "forms of pure intuition" (time and place),

but according to Kant's (at bottom relatively empiricist) notion
of knowledge this is not true of noumena.

   According to Kant, basically we can only gain knowledge
about the real world by means of **sensible** intuition, viz. by
means of singular representations related to sensibility.[1] This
can be argued to be Kant's empiricist mistake. Why not allow for
singular representations that are only indirectly related to the
senses just as we allow our scientific theories to be only indirectly empirically testable? Indeed why not be a scientific
realist (employing theoretical terms analogous in function to
something like "non-sensible" intuitions could be thought to
have) instead of an empiricist (see below)?

2.   How is, or can, all this be connected to scientific realism?
The crux of the reply is that a scientific realist can replace
Kant's noumena by scientific objects, i.e. by objects postulated
by the best-explaining scientific theories. Apart from that a
scientific realist naturally ends up claiming that we can have
knowledge about these objects - Kant's empirical intuitions are
in a sense replaced by factual intuitions which do not confine
to mere sensible ones. Thus the Kantian categories and forms of
intuition (or rather, their counterparts) become in a sense
indirectly applicable to scientific objects via an analogy: the
best-explaining scientific theories will ultimately provide for
analogical counterparts to them, and, accordingly, scientific
objects become knowable to us. On the other hand scientific
objects are epistemically inexhaustible - they may be conceptually approached in an infinite variety of ways. The result is
a brand of scientific realism with a Kantian flavor. In fact we
shall come rather close to the train of thought presented in
Chapter 2, and to the general notion of scientific realism discussed there. For that view contained the idea of a stereoscopic
picture of the world. Kant's philosophical distinction between
the phenomenal and noumenal worlds plays largely the same role
as does the distinction between the manifest and the scientific
images of Chapter 2 (cf. especially the second option concerning
the relationship between these images in Section II of Chapter
2) - even though these two distinctions are not the same.

   To put it more precisely, we can pave the way to the brand
of scientific realism (greatly influenced by the philosophy of
Wilfrid Sellars) to be presented and defended in this book by

making two important modifications to Kant's system (cf. also
Rosenberg, 1980). The first one is precisely the substitution of
scientific objects for Kantian noumena. The real or "noumenal"
world which backs and supports the "phenomenal" world is not a
metaphysical world of unknowable things in themselves but the
world as construed by the ideal, best-explaining scientific theories. The second one is this. Kant's static a priori categories
are to be analyzed as changeable functional ones at bottom,
indeed as something linguistic or at least as something analogical to linguistic entities - in effect as (broadly) semantical
roles in something like the Sellarsian sense of Chapter 3 (see
also Sellars, 1967b, and 1968). Thus we end up maintaining that
the categories of thinking are developing (in the sense in which
language as a form of social practice changes and develops) and
social. Recall that Kant's philosophy notoriously lacked these
features: His system of categories and a priori principles
simply is incapable of properly processing feedback from experience.

As noted in Chapter 3, Sellarsian semantics adds to this
picture a further feature, viz. a nominalistic view of abstract
objects (such as redness or triangularity) and claims that such
allegedly abstract objects in fact are "reducible" to episodes
of "languaging", where only particulars (such as that red ball
or this triangle) are involved. Furthermore, Sellarsian semantics leads us to the view that to predicate, to assign a property to an object, is in fact to classify the object conceptually: the object is in a sense modified, or displayed as, an
object of a certain kind, and predicates are a kind of (objective) bases or principles of classification (cf. Chapter 3 and
e.g. Sellars, 1968, and especially 1979).

The second modification of course involves the rejection of
immutable transcendental a priori principles. Corresponding to
them we have - where the need arises - general factual assumptions, sometimes the axioms of a scientific theory (cf. methodological presuppositions, general principles of causality, foundations of physical geometry, the axioms of Newtonian mechanics).

According to the well-known "Copernican revolution" of Kant
- to the extent it has succeeded - ontological categories are
epistemic at bottom. In Kant's words, "it should be possible to
have knowledge of objects a priori, determining something in
regard to them prior to their being given" and " we can know a

priori of things only what we ourselves put into them" (Kant, 1787, B xvi and xviii).[2] What there really is, then, is funda- mentally based on conceptual systems and on our knowledge (or capacity to know). (See Chapter 6, Sections I and II for an analysis and a discussion of the this kind of view.) In the case of scientific realism this partly comes to mean that what there really is in the world is decided by best-explaining scientific theory. Scientia mensura - science is the measure of everything (or at least of what really exists), we can say. I shall return to the epistemological nature of scientific realism in Chapter 6.

Let us still note that there is, in the Anglo-American literature on the philosophy of science, an ongoing lively debate about Putnam's (1978) distinction between metaphysical and internal realism. At bottom the distinction has to do with whether existence (ontology) is an epistemic or a non-epistemic matter. Internal realism corresponds precisely to the above- mentioned epistemic realism (or to one of its versions), whereas metaphysical realism takes ontology to be radically non-episte- mic (similarly the notion of a "really" existing object is non- epistemic). (I shall examine this distinction more closely in Chapter 6.)

II. General Arguments for Scientific Realism

1. Generally speaking, scientific realism is a doctrine which claims that the method of science is the criterion of what there is and isn't. Or, as the method of science is typically supposed by a realist ultimately to produce true theories, we may also say that, according to realism, science typically aims at or at least should aim at finding out what the world is like, viz., at finding true theories about the world (both about its observable and unobservable parts and aspects), and realists typically be- lieve that such theories can in principle be found (even if no unique true account of the world perhaps need be claimed to be attainable). Truth here involves - at least in part - specifying what the world is like, and this is typically supposed to re- quire some kind of correspondence idea of truth. (See Chapter 9 for a better and more naturalistic account of the aims of scien- ce and scientists.)

So for realism truth can be regarded as a kind of criterion

for the success of science, even if in actual practice at best approximately true theories can be hoped for (if we are to rely on the history of science at all). Let it be remarked here that van Fraassen in his recent book defines scientific realism basically in this way - but with one addition. For he also claims that according to realism the "acceptance of a scientific theory involves the belief that it is true" (van Fraassen, 1980, p. 8). But - unless one stipulates that to accept a theory means to accept it as true - this is not an epistemic attitude a realist always needs to adopt, at least without qualifications. I would like to say that according to realism a scientist only needs to accept as true (or approximately true) theories which indeed are true (or approximately true). Thus a realistically minded scientist can very well accept a theory as an instrument if he finds it useful for achieving his goals (say, prediction and control of observable phenomena).

In this book we shall defend only internal or epistemic scientific realism which rejects the Myth of the Given. As primary rivals of this kind of realism we regard metaphysical realism and instrumentalism, especially instrumentalism based on foundational empiricism. As other rivals of scientific realism we have (ontological) idealism and - perhaps - some "constructivist" versions of pragmatism. (Kuhn, Hanson and, perhaps, Rorty might be mentioned as examples of such constructivists.) Idealism in the sense of phenomenalism ("the world is phenomenal") will not be my main opponent in this book, as I believe it has been convincingly argued to be wrong (cf., e.g., Sellars', 1963a, and others' - by now - standard rebuttals). In any case, we shall below concentrate on defending internal scientific realism primarily against empiricist instrumentalism (including, e.g., van Fraassen's (1980) "constructive empiricism") and metaphysical realism. (Metaphysical realism is supposed to accept the Myth of the Given and we shall leave our comments on it to Chapter 6.)

Instrumentalism can be regarded as a view according to which scientific theories are merely instruments for the systematization of empirical (or observational) statements and for making empirical predictions. Standard instrumentalism also claims that the theoretical terms of scientific theories either do not (and cannot) succeed in referring at all or succeed in referring at most to empirical entities. Here 'empirical' means observable in a broad sense. Standard (empiricist) instrumentalism regards the observable/unobservable dichotomy as an **onto-logical** one and claims that, indeed, there are no unobservable entities at all. More sophisticated versions of instrumentalism may, however, accept the existence of at least some unobservable entities. An example of this kind of doctrine is provided by van Fraassen's constructive empiricism according to which "science aims to give us theories which are empirically adequate and acceptance of a theory involves the belief only that it is empirically adequate", viz. the belief that what the theory says about what is observable (by us) is true (see van Fraassen, 1980, pp. 12, 18). But my interpretation is that ontologically van Fraassen's constructive empiricism - on pain of contradiction in terms - is realistic (see, e.g., van Fraassen, 1980, pp. 31, 55, 64).[3]

2. We can now begin our examination of the arguments for realism. We shall contrast the manifest image of the world or the empiricists' "everyday" world with the scientific image of the world produced by science (cf. Chapter 2). As a well-known example let us have Eddington's tables. The table of the manifest image is, e.g., a table characterized with the help of observational predicates (a table as a kind of brown, hard-surfaced, wooden thing designed for human purposes), while the table of the scientific image is some kind of a conglomerate of subatomic particles. Which one of these ways to conceptualize the table is the "right" one? Are these concepts of table in some sense contradictory? Do we even have, in some sense, two tables (or only one)?

Empiricist instrumentalism says that only the observational table exists - microphysical theoretical predicates do not refer to any real entities. Scientific realism in its **strong** form (cf. option (2) in section II of Chapter 2) says, to use Sellars' jargon, that although the observational table exists relative to the manifest image, it does not **really** exist, for no single entity (such as a whole or system) in the scientific image corresponds to the observational table, only a conglomerate of objects (cf. Sellars, 1968, pp. 149 - 150). Consequently it is possible to associate with scientific realism the view that the two concepts are incompatible in the indicated sense, to yield strong scientific realism. According to **weak** scientific realism - which we adopt - the table as described in terms of either the

manifest or the scientific image can be taken to exist in both
cases, although the scientific description of the table is more
accurate and truthlike. In general, whether the objects and
features of the manifest image can be taken to exist depends on
how accurate and truthlike the best accounts within the manifest
image are, no matter whether the scientific counterpart notions
for such single objects represent wholes, systems or mere
classes of entities. If for instance, such best accounts (theories and singular descriptions) have true or approximately true
counterparts (not necessarily about wholes, as Sellars requires)
within the scientific image, then we can take the objects and
properties they speak about to exist (cf. note 2) of Chapter 9
on counterparts). (A case in point would be the existence of
mental states as conceived by best-explaining psychological
theories within the manifest image versus future best-explaining
theories covering them, in terms of counterparts or something
analogous, within the scientific image.) If the arguments below
will carry weight, our claims about the explanatory inferiority
(and this sense ontological inferiority) of the manifest image
relative to the scientific image should be warranted.

Let me now go into my present general philosophical de-
fenses of scientific realism. These arguments - or argument
sketches - agree in claiming that scientific realism takes
science seriously. The most basic line of argumentation starts
from the premiss that the scientific method is the best (e.g.,
most valid and reliable) method for achieving knowledge about
the world (cf. our discussion of this in Chapter 10). To this
premiss it adds another, controversial one claiming that when
the method of science is considered and applied in all its
richness - as happens within the scientific image - unobservable
explanatory entities are freely postulated in order to produce
true or approximately true accounts of the world. In other
words, according to realism the methods of science presuppose
realism; and this is what we will concentrate on below. Note
that an empiricist instrumentalist (in the traditional, strict
sense) cannot postulate explanatory unobservable entities (except as fictions). A constructive empiricist such as van Fraassen (1980) is willing to accept the existence of unobservable
scientific entities - and indeed is an ontological realist. What
is common to all empiricists and instrumentalists - or at least
to the positions commonly discussed in the philosophical literature - is that according to them the aim of science is to pro-

duce only empirically adequate accounts - theories - of the
world rather than to make factual truth (or truth about the
whole world) the criterion of the success of science.

The realist's basic claim then is that the scientific
method (in its fullest sense) at least in principle can lead to
best-explaining theories of the world (and that nothing else
can); and as ideal, best-explaining theories are assumed by the
epistemic realist to amount to true theories, the scientific
method can lead to true theories (and that nothing else can). A
weaker relevant thesis - which in fact suffices for our purposes - is that unless the scientific method is employed in its
fullest (viz. realist) sense it cannot lead to true (best-
explaining) theories.

Viewed from a different angle, one can make a related point
about existence in Sellars' terms: "to have good reasons for
holding a theory is ipso facto to have good reason for holding
that the entities postulated by the theory exist" (Sellars,
1963a, p. 91). Without really trying to specify and defend this
here, we can in any case loosely say that, accordingly, to the
extent one has one of these reasons to about the same extent one
has the other one. The entities postulated by a theory can of
course be unobservable ones, viz. entities an instrumentalist
cannot take to exist.

Some authors distinguish between **realism about entities** and
realism about theories (cf. Cartwright, 1983, Hacking, 1983).
Realism about entities means the view that entities (such as
electrons) really exist, while realism about theories means that
there are (or can be) theories which are true in some kind of
correspondence sense. Cartwright, 1983, for one seems to claim
that there can be no strictly true scientific theories while the
entities spoken about in these theories may or do exist. But I
maintain, on the contrary, that realism about entities cannot be
true unless a relevant kind of realism about theories is true.
(The converse holds, of course, too.)

I shall not go into this matter in detail here for I have
basically justified my view already. That is, if the Myth of the
Given is wrong then there can be no a priori given world (viz.
no a priori given categorizations of entities) and no theory-
independent knowledge of such entities. While the word 'theory'
should be taken in a loose sense here, I still think that this
contradicts Cartwright's view or, anyway, the view I above
attributed to her. Of course, I do not claim that our present

scientific theories are true or (mostly) even approximately true - the question is whether there **can** be true theories, nor do I claim that the connection between reference and truth should be made as tight as in Tarskian semantics (where satisfaction and reference are converse relations). And note, too, that it is not necessary to construe theories linguistically or by means of formal languages, for one can accept e.g. a structuralist view instead. Hence Gödel's incompleteness results and other similar results concerning the incompleteness of linguistic machinery - syntax - have no bite here.

To give a full account of the conditions under which the scientific method as a matter of fact leads to best-explaining theories (or theories "worth holding") is a major problem, however; but at least in some simple cases we have an idea of how to solve that problem (see Chapter 9 below). Irrespective of that, what is central for us presently is the realist's claim that unless the scientific method is interpreted and applied in the broad realist sense science cannot produce best-explaining theories and, accordingly, true or approximately true theories. In other, and more general words, science cannot perform the task of telling what the world is like unless it (or the scientific method) is interpreted realistically - and, it is added, no other method can carry out that task.

For the above general basic argument for realism to be acceptable to an instrumentalist (or, more generally, to a non-realist) best explanation presumably has to be restricted to observable phenomena and thus to exclude unobservable explananda. Of course a realist does not and should not concede this much to his opponent, for he thinks that any aspect or part of the world, be it observable or not, is equally well an object of theorizing and often (but not always) in need of deeper explanation. But even if he would do so if only for the sake of argument, he is able to win the battle against the instrumentalist, I claim. Why? There are several answers to this, and I shall now start discussing them. I shall accordingly give some arguments for the claim that the scientific method cannot lead to truth unless interpreted realistically.

First, it is a brute historico-sociological fact about scientific research that scientists have postulated unobservable entities such as electrons, black holes, viruses, genes, etc. and have been able to build successful explanatory theories employing concepts for such entities. Even if these theories have not perhaps been best-explaining theories and have been - or will be - replaced by better ones, a non-realist (instrumentalist or constructivist) is at a loss in explaining the "observable" fact that science has been remarkably successful. For he must at the same time explain away the most natural, viz. realist interpretation of science, and claim that scientists have somehow been deluded when postulating unobservable entities. It is plausible to think that a non-realist cannot succeed in this, for science does indeed confirm the evolutionary, naturalistic view of the world of philosophical realism (including the idea of human beings striving for correct cognitive representations of it). Put briefly and idealizingly, actual science typically is (or is like) what realism says it is (and should be). For a realist there is of course nothing to account for - at least if he is able to deal with transtheoretic reference, perhaps by means of some principle of "charity" and by assuming that the success of scientific theories must be measured from the point of view of their successor theories (cf. Putnam, 1981, Boyd, 1983).[4]

Secondly, a realist (typically) is a naturalist and emphasizes that humans are in many ways limited and restricted beings of nature. Thus, as is well documented, human senses are in some special circumstances rather defective and unreliable (even if they function well as information gathering-devices in normal circumstances). Thus it seems quite unwarranted to rely on the senses (e.g., on observation) in ontological and epistemic matters. What if our senses would change radically due to, e.g., mutation? The senses cannot be a (philosophical) criterion of what there is in the world. Yet that is what empiricism basically assumes. (The present argument of course does not apply against other non-realists than empiricists.)

Thirdly, according to realism, the manifest image is inadequate (in the sense of incompleteness) for successful explanatory theorizing about the world. Here we may think of the framework of the manifest image as something with ontological commitments - for that is how an empiricist construes it - or we may think of it as a conceptual framework together with some fundamental methodological postulates (as we did in Chapter 2).

This third subargument has several faces. I will below discuss two of them and consider the world as conceived within the manifest image from the point of view of a) its **"unstability"** (lack of nomicity) as well as b) its **explanatory incom-**

pleteness. The explanatory incompleteness of the manifest image (argument b)), when this image is construed ontologically, is a broad topic which will occupy us at length not only below but also in most of the next chapter.

3. To start our discussion of the explanatory incompleteness of the manifest image we recall that empiricist instrumentalism accepts the Myth of the Given - or in any case our main target in this subsection will be the kind of instrumentalism which accepts it. Let us recall the main versions of this Myth. The ontological Myth (MG_O) says that the world is ontologically ready-made. Applied to empiricism this means that the world is as our senses register in ideal conditions. The epistemic Myth (MG_e) in turn holds that there are self-justifying mental states and therefore knowledge which is in no way based on other knowledge (or "preknowledge"). Foundational empiricism typically assumes that there are states of awareness which are "ostensively" tied to the empirical world so as to give, in principle, incorrigibly certain knowledge of it. On the semantic or linguistic level the Myth of the Given (MG_l) means that the relationships between language and the world are necessary, ostensive ties. In other words, it is thought (e.g., in concept empiricism) that there are fixed and immutable semantic language-world relationships. The supporters of the Myth of the Given may deny subscribing to any of these versions; nevertheless philosophical analysis can reveal that they rely in their thinking on it.

One crucial ingredient in the argument under discussion is the attempt to show that the conceptual framework of the manifest image or "everyday" thinking is in a sense incomplete: the conceptual equipment of the manifest image is incapable of representing and sustaining stable general claims, that is, nomic generalizations. I shall illustrate the matter by using an invented (though often discussed) example. Let us suppose that gold dissolves in aqua regia sometimes with rate n_1, sometimes with rate n_2, and that no observational properties of either gold or the circumstances have been found (even after arbitrarily thorough examination) to "explain" this variation or the unsatisfactoriness of the empirical generalization. On the other hand the best available chemical theory can explain these matters by assuming that gold can have two distinct types of underlying microstructure, call them m_1 and m_2. In terms of the scientific image, the dissolution rate of gold with microstructure m_1 is n_1 and that of gold with microstructure m_2 is n_2. When the total situation is given a new description in terms of the scientific image, the unstable and non-nomic feature of the manifest image, viz., the variation in the dissolution rate of gold, is explained. (This example has been analyzed in more detail in Tuomela, 1973, Chapter VII; also see below.)[5]

There is an argument for realism which bears a resemblance to the one just given, namely, the so-called "miracle argument". According to it the stability and completeness of the manifest image presupposes a kind of miracle or "cosmic coincidence". For instance, if there were no electrons, it would be a miracle if certain phenomena took place in galvanometers and cloud chambers. Thus instrumentalism which relies on the manifest image cannot always causally explain interconnections between observational phenomena. We can then claim that the theoretical terms of theories in fact manage to refer: among other things they may refer to such common causes. Such inference to best explanation (in terms of unobservables) in certain important situations is central for certain brands of scientific realism, but it is of course unacceptable to empiricists. How to restrict appropriately the scope of inference to best explanation is a difficult problem we cannot address here - certainly this principle does not apply mechanically to all cases involving common causes (contrary to what e.g., van Fraassen, 1980, p. 23ff. claims about realism). A case in point is the problem of the existence of so-called hidden variables in the realm of microphysical phenomena - see the appendix below for a discussion of them and the alleged threat for realism posed by investigations related to Bell's inequalities. In general, I would like to say that it is the scientific serendipity involved in finding best-explaining theories about the world that is central - from where, if anything at all, such best-explaining theories are inferred is a side issue or at least not the key issue.

Empiricist instrumentalism nevertheless claims, precisely on the basis of the Myth of the Given, that the manifest image is stable. For an empiricist the limits of the world and the limits of (actual and possible) observations in principle coincide. In this respect the world is identical with the world as a perceptual object, and this latter in turn is assumed to be ontologically given and ready-made. The epistemic and linguistic versions of the Myth of the Given then guarantee that the world is in principle cognizable and representable in symbolic form. The empiricist's world is closed with respect to observations and it is an a priori stable (or law-sustaining) totality. Accordingly, the empiricist cannot accept the claim in the above example according to which the observational anomalies, coincidences, etc., are in principle unexplainable within the observational framework. Put briefly, for an empiricist the empirical (or observable) world is explanatorily closed and lawful, whereas for a realist it is not. According to realism there can be lawful observational generalizations at most when scientific theory is suitably injected into the manifest image (see the above example).

I shall now formulate, on the basis of what was said, a simple, logically valid deductive argument for realism, using the word 'true' in a wide sense:

(**WR1**)(P1) If (empiricist) instrumentalism is not true, then scientific realism is true.

   (P2) If the world conceptualized in terms of the manifest image is (in principle) stable on a priori grounds, then the Myth of the Given is true.

   (P3) If the world conceptualized in terms of the manifest image is not stable on a priori grounds, then instrumentalism is false.

   (P4) The Myth of the Given is false.

   (C) Therefore, scientific realism is true.

The premisses (P1) - (P4) are assumed to be true on philosophical or conceptual grounds (i.e. they are assumed to be non-contingent). The thesis (P1), to be understood as making ontological commitments, says that if instrumentalism is an incorrect doctrine, realism is a correct one. Instrumentalism is here understood as claiming that theoretical scientific predicates do not refer to anything - or not to unobservable objects at any rate. Thesis (P1) seems to hold true at least if realism is held to maintain no more than that at least scientific objects (e.g., Eddington's scientific table) exist - given that there are no non-scientific objects such as demons, spirits and the like. We are making this assumption in this work (see Chapter 7, Section II). We have called this variety of realism **weak** scientific realism.

If, on the other hand, realism is understood to be **strong**
or **exclusive** realism according to which only scientific objects
exist (and according to which the observational table does not
really exist), the truth of (P1) is very problematic, to put it
mildly. For the claim that instrumentalism is false, involving
the thesis that scientific theoretical predicates do succeed in
referring to something, holds even if also empirical objects
(such as observational tables) are taken to be real, and so weak
realism makes (P1) true. And since we have already at the outset
of our discussion excluded the possibility of the existence of
non-scientific objects (e.g. demons), realism can and should be
taken to mean nothing more or less than weak realism.

How about (P2)? I think it is obvious that the a priori
stability of the manifest image is essentially connected with
the Myth of the Given (to its epistemic version, in the first
place), though I am not quite sure if the relation is just that
the Myth of the Given is, strictly speaking (without further
assumptions) a necessary condition of the a priori stability of
the world. I am willing, nevertheless, to accept (P2) on the
basis of the above discussion. (Note that if, after all, the
manifest image turned out to be lawful a posteriori, on the
basis of scientific research, that would not affect (P2).)

And (P3)? It has it that instrumentalism cannot be correct
unless the manifest image is a priori stable (law-sustaining).
For otherwise instrumentalism would concede that there are no
conceptual or philosophical grounds for its claim that it can
give an adequate explanation of its intended explananda (viz.
empirical phenomena and structures). We could, though, argue
over the epistemic and metaphysical status of (P3), but I shall
not go into that here. The least we can say is that the falsity
of (P3) would make instrumentalism difficult to defend. (Let us
also note that the converse claim that stability entails instrumentalism, too, can be regarded as true, at least if we accept
also certain other philosophical assumptions of Sellars. Sellars, 1963a, p. 96, in fact relies in his defense of realism
mainly on this converse entailment.)

Premiss (P4) has been argued for in Chapter 3. I shall here
merely point out that it is a very central assumption, at least
for epistemic scientific realism. Its positive content is, in
part, that there is no a priori privileged conceptual scheme for
describing the world and gathering information about it. If, for
instance the empiricist Myth of the Given would be true, our

everyday language would bear irreplaceable and necessary ties to empirical reality. It would then be conceptually impossible to replace it by some other, e.g. a scientific, language. Rejecting the Myth of the Given makes it also possible that theoretical terms can have an "evidential role", i.e. that they can in principle occur in singular claims which are used in direct description of the world (see Sellars, 1967a, p. 352 and Tuomela, 1973, Chapter V).

A somewhat weaker, pragmatic version of the discussed argument is obtained by using the phrase 'It is reasonable to accept that' in front of each of the four premises (P1), (P2), (P3), and (P4) and the conclusion (C). This new pragmatic argument can be regarded as (conceptually) valid because the original is and because of the logical behavior of the notion of reasonable acceptance.

The rejection of the Myth of the Given can accordingly be said to give a good reason for rejecting the empiricist's claim that observation is all that matters. According to the realist who rejects the Myth of the Given, theoretical terms can **in principle** be used in direct reporting in an evidential sense and thus they are not parasitic on observational terms. Thus the empiricist's claim that empirically (observationally) equivalent theories are always equivalent has not only been shown to be false (briefly and trivially: they need not be equivalent, as their theoretical parts may make different claims about the world and attribute different ontological structures to the world) but philosophical grounds for its falsity have been provided. (See, e.g. Tuomela, 1973, pp. 58 f., 116 - 120, van Fraassen, 1980, and Boyd, 1983, for a relevant discussion of the thesis that empirically equivalent theories are evidentially indistinguishable.)

A related matter, indeed the other side of the coin, is that an empirical generalization can be made stable (cf. the antecedent of (P2) above) only - and at best - if **theory is injected into it,** viz. only if it is laden with an explanatory theory. Put in terms of our above **aqua regia** -example, we can get a stable empirical generalization only if the information about the relevant microstructures m_1 and m_2 can somehow be incorporated in the description - in terms acceptable to an empiricist. Although there are technical devices, such as the Ramsey-sentence method, which have been proposed for more or less this kind of eliminative purpose such approaches have not

been very successful - in the first place because the predicate-
variables of Ramsey-sentences have not been shown to have observational substituends as required by the empiricist programme.
(See our discussion in Chapter 5 below and in Tuomela, 1973).

    We have above concentrated on arguing generally against the
versions of empiricism and instrumentalism which accept the Myth
of the Given. In the next chapter we shall continue our discussion concerning arguments for realism and concentrate on more
specific, technical arguments which do not directly rely on the
rejection on the Myth of the Given and which go against all
versions of ontological and methodological instrumentalism and,
more generally, non-realism, including constructivism. It is
obvious that internal (or epistemic) realism and constructivism
(cf. Kuhn) bear many similarities in emphasizing the importance
of social and conceptual factors in the scientific enterprise.
Our main criticism against constructivism is its lack of naturalism and the ensuing inability to explain the transtheoretic
reference of theoretical terms and the (continuing) success of
science (cf. Boyd, 1983, and our naturalist, causal account of
picturing in Chapter 6). And note that our direct arguments for
realism, if sound, go against all forms of non-realism, anyway.

    It is of interest to consider also arguments against a
given position and not only for it. Thus, if there are devastat-
ing criticisms against scientific realism it must of course be
abandoned. And, indeed, some theoreticians have thought that
some arguments due to the situation in quantum mechanics show
that realism is untenable. I am here referring to the situation
created by the experimental refutation of Bell's inequalities.
My view is that nothing important concerning the tenability of
scientific realism follows from those results. My arguments will
be presented below. Because of the somewhat special and technical nature of this material it has been presented in the form of
an appendix.

APPENDIX ON QUANTUM MECHANICS, BELL'S INEQUALITIES, AND SCIENTIFIC REALISM

1. The philosophical interpretation of quantum mechanics has
been a widely debated topic since the very birth of this theory
in the 1920's. To mention just one prominent issue, Einstein,
Podolsky, and Rosen (1935) presented a famous thought experi-

ment, which was assumed to show that quantum theory is in a sense incomplete. The debate between Einstein and Bohr related to this issue has become a classic on this topic. On the whole we can say that the (or at least some) basic issues involved seem to be realism (or, rather, the viability of hidden variable theories) and locality as well as, to a less extent, determinism (whatever these notions exactly are taken to involve).

Much heat to this discussion was recently added by Bell, who proved an interesting result called "Bell's theorem" or "Bell's inequalities" (cf. Bell, 1964, 1971). These inequalities are fairly well empirically testable and indeed they have been tested and seem to come out as false. This fact has lead some investigators to claim that (scientific) realism is thereby refuted. As the tenability of realism is very much the topic of this book, it is certainly of interest if the best current microphysics could be taken to entail the unacceptability of realism. Let us take a closer look at this complex and intricate matter.

The experimental situation most commonly discussed in connection with Bell's inequalities is the so-called spin correlation experiment. In such an experiment we are concerned with a system in which two spin-1/2 "particles" or "quantons" are emitted so that the spin of the total two-particle system is zero and the particles are detected in a Stern-Gerlach type experimental setting with two detectors. The particles move in different (often strictly opposite) directions and the spin of each particle (e.g. electron) is measured as it reaches the detector. One interesting thing about the system is that the quantity of spin and also the spin component of the total spin in any given direction must remain the same throughout its evolution. Thus if one particle has the spin component value +1/2 the other one must have -1/2, this conservation principle says. (If our particles are photons, as we shall typically assume below, their spin component values are +1 and -1, also summing up to zero.) However, the particles are spatially separated and so the observed existence of such correlation may seem puzzling, if there cannot be action at a distance. It is not surprising, therefore, that theoreticians with strong causal intuitions have e.g. postulated "hidden variables" as a kind of common causes to account for such puzzling correlations. Bell's inequalities and their experimental testing has served to sharpen the discussion of such common causes.

The general methodological situation here is as follows. We
are dealing with a (possible or actually existing) microphysical
theory, say T. T may be current quantum theory or it may be some
kind of hidden variable theory. T is in any case supposed deal
with microphysical phenomena in an informative way. However, as
is typically the case in science, T does not by itself yield
experimentally testable predictions. But it is still (or at
least should be) a testable theory, and hence conjoined with
suitable auxiliary assumptions it will have empirical, and indeed experimental, import. Let now A be the conjunction of such
relevant auxiliary assumptions. Then the conjunction T&A will
(deductively or inductively) imply experimental statements, say
K, and this we denote by T&A → K.

In the case of Bell's inequalities, T is typically a hidden
variable theory and T (or at least T&A) normally contains some
kind of conservation principle, a locality principle and in some
derivations (although not in Bell's own) even some kind of
"realism" principle; the arrow denotes deductive implication. As
will be seen below, in this situation K will concern experimental detection probabilities (or expected values of random variables). These probabilities typically concern the probability of
detecting particles with certain spin component values.

Hidden variable theories have a relatively long history
(see Jammer, 1974). As said, the general idea here has been to
look for some kind of common causes for the underlying quantum
phenomena. In the case of Bell's inequalities we can say that
historically the search for common causes has proceeded from
strictly deterministic causes to increasingly less strict probabilistic ones. Specifically we can say the following about
testing Bell's inequalities. By assuming a kind of hidden variables one can derive experimentally testable inequalities described by K. On the contrary, quantum mechanics implies for
those situations certain other results, say K', such that K' →
~K, viz. K and K' are logically incompatible. According to
recent experimental investigations K' agrees with the verdict of
the world while K does not (see, e.g., Clauser and Shimony,
1978, for a survey). As the implication T&A → K can be written
in the logically equivalent form ~K → ~(T&A), we are now allowed
to conclude that ~(T&A) is true. So the problem is to decide
whether the truth of that negation should be brought about by
assuming ~T (viz. that the hidden variable theory T is false) or
by assuming ~A (or by making both assumptions at the same time).

In the earlier investigations related to Bell's inequalities **deterministic** common causes were looked for, viz. such hidden variables as jointly with information concerning the polarization of the measurement devices strictly determine the spin components (or at least the spin-measurement results) of both particles.[6] In the case of such deterministic hidden variables it is assumed that the theory entails a strict (non-probabilistic) condition of **separability** (locality), according to which - after it has been "operationalized" to fit the experimental situation at hand - the results of the measurements of the spins of the two particles are independent of each other. Let us formulate this important separability principle in precise mathematical terms. Thus denote by A_a the result of a measurement of the spin component of the left particle in a standard spin correlation experiment when the direction of polarization is a and the hidden variable is denoted by x; and let B_b be that for the right particle, with the direction b. Then the separability assumption amounts to this:

(1) $(A_aB_b)(x) = A_a(x)B_b(x).$

The hidden variable theory in question is assumed to include the specific assumption of the conservation (anticorrelation) of the spin component values. The auxiliary assumptions include a variety of background assumptions concerning the proper operation of the measurement apparatus. We also have here the general underlying assumption that standard Hilbert space mathematics on the whole is applicable to the study of microphysical systems. (This is of course an assumption shared by quantum theory as well.)

In the case of deterministic hidden variables the principles of separability (or locality) and anticorrelation (or conservation) essentially suffice for the derivation of Bell's inequalities. Actually we may separate our present assumption of deterministic locality into two subassumptions, viz. into that of determination and the above separability assumption. Determination involves that x together with the direction of polarization determines the values of the variables A_a and B_b. Accordingly also the product (A_aB_b) can be assumed to have a uniquely determined value.

As Bell's inequalities can be derived also from weaker assumptions we shall not dwell on the above deterministic case, which also seems too strict in view of experimental results (cf. Clauser and Shimony, 1978). Let us first note that the above notion of separability can be relaxed for the case of probabilistic hidden variables to make possible an investigation of probabilistic common causes, viz. common causes which together with direction of polarization make spin-measurement results probabilistically independent. In this case we think of the variables A_a and B_b as random variables in the standard sense. Locality in the sense of **probabilistic separability** now means that the expected values of these random variables with respect to the hidden variable are independent, viz.:

$$(2) \quad E(A_a B_b/x) = E(A_a/x)E(B_b/x),$$

where E means expected value.

Analogously with the deterministic case but using probabilistic separability and omitting the determination requirement we may now derive the following inequality as Bell did (1971) (also cf. Clauser and Shimony, 1978, p. 1893):

$$(3) \quad -2 \leq E(a,b)-E(a,b')+E(a',b)+E(a',b') \leq 2 \ ,$$

where a' and b' refer to alternative directions of polarizations and where the expected values are shorthands such that, e.g.,

$$(4) \quad E(a,b) = E(A_a,B_b)$$

such that

$$(5) \quad E(A_a,B_b) = \int_x (A_a B_b \ dx).$$

It may be noted that even if Bell's treatment of the probabilistic case does not assume the mentioned kind of strict determination it still comes close to it. For if we in addition assume that

$$(6) \quad E(A_a/a,b,x) = E(A_a/a,x), \quad \text{and}$$
$$E(B_b/a,b,x) = E(B_b,b,x),$$

viz. the probabilistic irrelevance of the measurements with respect to the other direction of polarization, then determinism follows (that is, the conditional variance of the spin measurement, given x and direction, is zero).

Bell-type inequalities have been derived also from somewhat different premises (see, e.g., Clauser and Shimony, 1978). But it can presumably be said that Bell's above derivation is based on the **weakest** and **least problematic** assumptions that have been used. For instance, Suppes and Zanotti (1980) need in addition a problematic probabilistic symmetry assumption, Eberhard (1977) a problematic modality assumption, and so on. Whatever other assumptions have been used, locality has generally been explicated as separability, as above. We recall that in the case of probabilistic common causes it amounts to conditional probabilistic independence (see (2)). Thus the common causes sought after in this type of investigation have been assumed to be this strong.

The experimental investigations testing Bell's inequalities (such as (3)) have come out against them (see, e.g., Clauser and Shimony, 1978, and Aspect et al., 1982). They are empirically false according to the best scientific knowledge of today, but, on the contrary, the predictions by quantum mechanics agree with the experimental data. In this sense the verdict of the world seems rather clear. What can we and should we conclude from this? As the anticorrelation principle assumed in Bell's derivation seems acceptable, we seem to be left only with the choice of rejecting separability. This condition is too strong (cf. our above comments). Microphysical systems of the type considered in spin correlation experiments simply do not seem to be separable in the sense (2).

The experimental refutation of Bell's inequalities does not warrant stronger conclusions than the above. Yet, in the literature one can find claims (by serious investigators) amounting to much more. One such claim - and a misleading one - is that "it can now be asserted with reasonable confidence that either the thesis of realism or that of locality must be abandoned" (Clauser and Shimony, 1978, p. 1883, for a similar view see also d'Espagnat, 1979). This claim is (at best) misleading because it directly involves the issue of realism in the matter. For we recall that nowhere in Bell's derivation was the assumption made that "external reality is assumed to exist and have definite properties, whether or not they are observed by someone", which is how Clauser and Shimony (1978), p. 1883, define realism. Thus, if we keep to Bell's derivation, the choice is not between realism and locality but rather between anticorrelation and locality in the separability sense - and the choice is not difficult to make.

One reason why physicists and philosophers of physics seem
so eager to involve the issue of realism here seems to be that
they have their traditional ontological way of interpreting the
hidden variables (our x above). That is, they seem to think that
hidden variables can be interpreted according to naive realism
so as to straightforwardly represent something in the world. But
it must be emphasized that this is an extra assumption one is
not committed to in the case of Bell's argument. We can consider
any physical theory as a bunch of mathematical axioms (or structures, if you prefer the structuralist approach) independently
of whether those terms somehow represent anything in physical
reality. What they so represent, if anything, is a different
matter and finding out that requires additional semantical assumptions (e.g. to the effect that a certain symbol refers to
such and such specific thing or state in physical reality).

It is important to notice that this applies to quantum
theory as well as to hidden variable theories: the problem of
semantical interpretation must be faced in the case of any
physical theory. Thus one need not and cannot a priori stick to
a certain way, e.g. the naive realist's or the instrumentalist's
or an idealist's, of interpreting physical theories. What is
more relevant to realism, one can well be a realist without
taking all the extralogical symbols of a physical theory to
refer directly to any physical entity. Thus, for instance, one
can accept a weak realist interpretation of quantum mechanics
which allows that e.g. the symbol 'ψ' for the wave-function does
not itself directly represent anything in reality even if the
whole theory can be taken to successfully speak about mind-
independent quantons and their wonderful world. In all, the
sometimes made assumption that a hidden variable theory somehow
necessarily involves the assumption of the existence of a mind-
independent real world (and perhaps specifically of the existence of determinate spin-component values of e.g. photons at all
times, independently of measurement procedures) is not correct.
It is not a priori necessary to interpret a certain language in
a given way - to assume that would involve the acceptance of the
Myth of the Given, criticized above in Chapter 3.

It is more correct to see the controversy between quantum
mechanics and hidden variable theories as follows. Quantum mechanics has the radical new feature that properties of physical
systems can be predicted completely only if the measurement

procedure is given (viz. only if the propensities and disposi-
tions of the relevant microsystems have been appropriately ac-
tualized) and these results are even then only probabilities.
The search for hidden variable theories has been motivated by
the desire to return to a theory of the type of classical
physics. In such a theory, there would be well defined physical
quantities obeying deterministic causal laws. The statistical
nature of microphysical phenomena would be due only to our
ignorance of the values of the hidden variables and our ability
to measure only certain averages. Thus a crucial point in the
search for hidden variable theories has been that they would
allow a realism of a certain definite narrow type, viz. that of
classical physics. If hidden variable theories would be shown in
general to be incompatible with observed phenomena, this would
show a return to that kind of realism to be impossible. So far
this has been accomplished only for strongly local (viz. separable) hidden variable theories. (There are also other reasons to
doubt a return to classical type of physics, but we shall not go
into them here.)

2. Bellian inequalities have also been derived for other than
hidden variable theories. One such derivation is due to Mittelstaedt and Stachow (1983). The special interest of their treatment for our concerns is that they specifically use a kind of
realism assumption in their derivation. Let us briefly consider
the matter. These authors' proof is based on two central extra-
logical premises. They are called the realism assumption (R) and
the locality assumption (L) and formulated as follows (Mittelstaedt and Stachow, 1983, pp. 519 - 520):

(R)   If the value of a physical magnitude A can be determined
       without in any way altering the physical system, S, then
       a property P which corresponds to the value of the magni-
       tude A pertains to the system S.

(L)   If two systems cannot interact with each other, then a
       measurement with respect to one system cannot alter the
       other system in any way.

     Mittelstaedt and Stachow show first that (R) and (L) con-
tradict quantum theory or, more exactly, the proposition that

the von Neumann-Dirac quantum theory is true. This simple proof
- which I will not repeat here - shows sharply the basic content
of the Einstein-Podolsky-Rosen "paradox" and indeed agrees with
the EPR conclusion that these assumptions are indeed mutually
inconsistent.

        Mittelstaedt and Stachow then go on to prove a simplified
version of Bell's inequalities on the basis of **(R)**, **(L)**, and the
anticorrelation principle. I shall not comment on that here
because they use the strong non-probabilistic locality principle
(L), which, by the way, is entailed by (but does not itself
entail) separability in the strict, non-probabilistic sense. But
it is interesting that these authors' proof uses the principle
(R), supposed to explicate realism in this context. Let us take
a closer look at this principle, due to Einstein.

        First of all, it should be said that we have two basically
different ways of understanding **(R)**, for we may regard it as
either an intra-linguistic statement or an ontological, language
- world statement. If we adopt the first interpretation, then a
separate account of ontological matters is surely needed and **(R)**
clearly falls short of what it is meant to accomplish in the
present context. It is undoubtedly the second interpretation we
must be interested in here, even if neither Einstein nor the
present two authors make the matter clear. So let us consider
the following:

(R*)   If realism is true then **(R)**.

Is **(R*)** tenable on either conceptual or metaphysical grounds?
Can we assume realism and the antecedent of **(R)** and yet allow
its consequent to be false possibly? We agreed that the antecedent of **(R)** refers to something linguistic, presumably a physical
theory, we may assume. Suppose now such a theory, say T, entails
the antecedent of **(R)**. But if T is false the antecedent may be
false as well. But it is obviously intended by the authors in
question that the antecedent be true and that the truth of the
consequent of **(R)** be grounded on this. Thus T must be assumed
true. But then, if we indeed take **(R)** to be a language - world
statement, its truth follows from the assumption of the truth of
T, for that truth must then be taken to be correspondence truth.
So the conclusion is that **(R*)** is indeed almost trivially true,
if **(R)** is interpreted as intended - which intended interpretation requires its antecedent to be entailed by a true (viz.

correspondence-true) theory. If again T is not true A cannot be taken to entail the consequent of (R), and (R*) is falsified. So in all, (R) (partially) explicates realism only when it is interpreted according to realism. But it could also be regarded as an intra-linguistic statement and interpreted e.g. according to pragmatism and by assuming assertability-truth so that no direct inference from its consequent to what the basic furniture of the world is becomes possible. In all, Mittelstaedt's and Stachow's claim that their derivation of Bell's inequalities must be taken to rest on the truth of realism is not acceptable, for although (R) of course may be interpreted realistically, one is not logically or semantically committed to it.

3. Our above investigations indicate that separability and other related strong forms of locality are not satisfied by micro-physical systems. But there are weaker types of locality which may be applicable to physical reality. Indeed, the notion of locality used in special relativity theory, viz. Einstein-local-ity, can be generalized to the probabilistic case so that it applies to the cases that Bell's inequalities are concerned with. Let us briefly consider the matter.

Einstein-locality means (strict or probabilistic) determi-nation in space-time such that the complete microstate of the backward light-cone of an event (strictly or probabilistically, respectively) determines the whole event. Put more exactly, let R be a region of space-time (viz. an event), let C_R be the backward light-cone of R, and let R' be another event with $C_{R'}$ as its backward light-cone. Let now P be a hidden variable theory together with all the needed background assumptions (e.g. P=T&A, in our earlier terminology). We may now consider the satisfaction of P by C_R and $C_{R'}$. Thus we define that P is probabilistically Einstein-local (relative to a given property, e.g. a spin component) if and only if it is entailed by the assumption that P is satisfied by C_R if and only if P is sat-isfied by $C_{R'}$ that according to P the probability of the prop-erty in question holding true of R equals the corresponding probability for R'. To put the same in somewhat sloppy terms, generalized Einstein-locality means that equivalent initial conditions C_R and $C_{R'}$ determine same probabilities for R and R'. (Cf. Hellman, 1982, for a relevant discussion and for a similar formulation of probabilistic Einstein-locality.)

We must leave a more precise treatment of the matter for another occasion, but the idea is, anyway, to generalize Montague's analysis of determination to the probabilistic case. Note that given this analysis the values of properties, such as spin-components, become determined locally within a certain space-time region. Thus this notion differs from separability, which is concerned with two (or more) different regions and their relation.

It is important to note that (probabilistic) Einstein-locality does not contradict quantum mechanics. For this notion is a generalization of the notion of locality employed by relativity theory, viz. just Einstein-locality, and relativistic quantum mechanics gives the same predictions for spin-correlation experiments as does non-relativistic quantum mechanics. (Cf. de Beauregard, 1983, also see Hellman's, 1982, similar claims and arguments.) Thus, even if standard quantum mechanics is non-local in the sense of separability (and e.g. the Einsteinian principle (L) above) it is still local in the sense of (probabilistic) Einstein-locality. Neither the assumption of the truth of quantum mechanics nor the assumption of Einstein-locality can be used in any important sense to derive Bell's inequalities, no matter what kind of other assumptions they are conjoined with.

4.   Our above investigations indicate that the search for hidden variable theories is or need not yet be ended. For there may be theories postulating common causes which are not in a strong sense (e.g. separability or (L)) local but only in some weaker sense and which in any case satisfy (probabilistic) Einstein-locality. Such a hidden variable theory would have to agree with quantum mechanics in its predictions for spin-correlation experiments but it might differ in some other respects.

No matter whether the story of hidden variable theories in microphysics will continue in some interesting sense let us recall that - if our argumentation has been sound - their existence is only indirectly connected with realism. Thus, in testing Bell's inequalities against experimental results we are by no means doing anything like "experimental metaphysics" - philosophy is not that simple and physics cannot do that much.

It is also relevant in this connection to notice that several notions of realism have been involved in the discussion

connected to Bell's inequalities. While I shall make no attempt
to disentangle them here, let me still briefly comment on the
issue. First, there is the classical idea of realism as the
doctrine that there is a mind-independent real world (see Section II of Chapter 6 for a clarification of mind-independence).
In the present discussion this classical sense gives a minimal
version of realism (the matter depends, however, on how mind-
independence is understood). As we noted in the beginning,
Shimony and Clauser (1978), for instance, add that the world has
"definite properties, whether or not they are observed by someone". D'Espagnat takes realism to be a doctrine according to
which "regularities in observed phenomena are caused by some
physical reality whose existence is independent of human obser-
vers" (d'Espagnat, 1979, p. 128). These two examples fall within
traditional realism, even if they are as such rather unsatisfac-
tory characterizations. (For one thing, they concentrate on the
unnecessarily strict notion of observation-independence instead
of the broader mind-independence.)

While the experimental refutation of Bell's inequalities is
not directly connected to the tenability of any kind of realism
there is, however, a version of realism which contradicts quantum mechanics and the results of spin-correlation experiments.
This realism, which might be called strict realism, accepts the
following principle of bivalence:

(SR)   The systems (correctly) described by quantum mechanics
        exist, and for any possible property of such a system it
        holds true that either the system has the property or the
        contrary of the property, independently of observation.

A property is here technically explicated as a subset of
the Borel set of the phase space associated with the system.
Thus, for instance, it is consistent with (SR) to claim of a
microphysical system that it always has a (pointlike) position
and momentum or, for that matter, spincomponent value. But this
claim we know to be false. (Rather we should treat such microphysical quantities as position and momentum as random variables
and thus without commitment to (SR).) In any case all forms of
realism accepting (SR) are too strong. (Thus also Clauser's and
Shimony's "definite properties" should not be understood in
accordance with (SR).) It is often thought that Einstein accepted a form of metaphysical realism embracing (SR). If so, he

was wrong. Quantum mechanics requires its own concepts, and
those concepts are quite different from concepts used by classical physics.

To end, let me emphasize the general philosophical point
that not only can microphysical theories such as quantum theory
be realistically interpreted, but they are universal in the
sense of being in principle applicable to all real systems, both
micro- and macro-systems as well as any systems falling in
between. As we know, to put the matter in vague and imprecise
terms, quantum mechanics is strictly speaking incompatible with
classical physics, but by treating Planck's constant as a variable and letting it approach zero - contrary to fact - we are
mathematically led to predictions coinciding in limit with classical ones.

While quantum theory is in principle a universal theory it
is still a methodological necessity to distinguish between mac-
ro- and micro-systems in actual scientific practice, where quantum mechanics is typically applied to micro-systems. Even if the
quantum world is sui generis and as such requires no reference
to other systems, still the measurement of quantum phenomena requires reference to macrosystems somewhat on the lines Bohr
thought. But it is a far cry that this methodological practice
entails that the quantum world somehow depends on the human mind
or even human observation and consciousness as some stricter
advocates of the Copenhagen school have claimed.

CHAPTER 5

METHODOLOGICAL ARGUMENTS FOR SCIENTIFIC REALISM

I. The Theoretician's Dilemma and Scientific Realism

1. Scientific realism can also be supported by arguments which are more indirect but at the same time somewhat more specific and detailed than those discussed in the previous chapter. Thus we can and will below present several methodological (and "inductive") grounds for the need and usefulness of theoretical concepts. In all, they give good grounds for refuting the so-called theoretician's dilemma, a dilemma which e.g. Cornman (1975) repeatedly used to defend instrumentalism.

The simplest form of the theoretician's dilemma argument runs as follows. Either theoretical concepts in science serve their purpose or they don't. If they don't, they are clearly dispensable. If they do, their work can, however, be done without them. So even in this case they are dispensable.

The theoretician's dilemma can be formulated in somewhat more detail as follows:

(TDA)(1) The theoretical terms of scientific theories either serve their purpose or they don't.

(2) If they don't serve their purpose, they are unnecessary.

(3) If they do serve their purpose, they create epistemic connections between observational claims.

(4) If they create such connections, these very same connections can be created without theoretical terms.

(5) If these connections can be created without theoretical terms, these terms are unnecessary.

(6) Therefore, theoretical terms are unnecessary.

(7) If theoretical terms are unnecessary, there are no reasons to postulate theoretical entities as their referents.

(8) If there are no reasons to postulate such theoretical entities, there are no reasons to believe that they

exist, i.e. ontological instrumentalism is a reasonable option.

(9)   Therefore, there are good reasons to accept ontological instrumentalism.

The intermediate conclusion (6) of this argument can be interpreted as a claim to the effect that epistemic (and perhaps also
linguistic) instrumentalism is acceptable - if understood in the
way specified in the premisses. The final conclusion (9) of the
argument then claims that even ontological instrumentalism is
acceptable.

There are, however, several good reasons against accepting
the theoretician's dilemma argument as sound. As I have elsewhere (in, e.g., Tuomela, 1973, 1978a) argued, its premises (4)
and (5) can be regarded as unacceptable - and this fact supports
scientific realism. To show that in detail would require more
space than available here. Let me, however, briefly discuss the
matter in the following.

Premises (1) and (2) of (**TDA**) are clearly acceptable. I
also regard (3) as acceptable: in one way or other theoretical
terms should be epistemically relevant to describing the empirical (observable) world, even if that connection might be indirect (theory-mediated). While a hard-core realist need not perhaps accept (3), we shall here do that for the sake of argument.

As Craig's and Ramsey's well known elimination methods
show, the deductive logical relationships a theory establishes
among empirical statements can technically be preserved by a
theory (a subtheory or Ramsey-sentence) couched only in empirical or observational terms. (The ontological commitments of the
Craigian and Ramseyan transcriptions of theories is a somewhat
tricky issue. I shall not here discuss the matter but we shall
assume here, for the sake of argument, in the standard fashion
that they indeed commit their users to observable entities but
not to unobservable ones.) We shall below regard theories as
conjunctions (or sets) of nomological general statements in a
pragmatically adequate scientific language, where pragmatic
adequacy entails the availability of pragmatically grounded
metalinguistic rules of inference concerning the predicates of
the theory (cf., e.g., Tuomela, 1973, Chapter V).

Thus we let $T(\lambda \cup \mu)$ be a scientific theory with $\lambda \cup \mu$ as its
set of extralogical predicates, where λ is the set of observational or empirical predicates and μ the set of theoretical

ones, relative to this theory. (It is not necessary here to worry how exactly the observational-theoretical dichotomy can or should be drawn. In any case, the distinction - for a realist at least - is not an ontological one but a pragmatic or methodological one related, e.g., to the limitations of the perceiving apparatus of human beings; cf. Tuomela, 1973, Chapter I.) Note, too, that observational predicates can accordingly be used to speak indirectly about unobservable entities, e.g. "There are entities which cannot be observed" or $(Ex)\sim(O(x))$, taking O to mean "observable"). For an empiricist it is necessary that there be a distinction between observational and theoretical predicates - otherwise his position would not really be distinct from realism. (This remark is especially pertinent in the case of van Fraassen's (1980) constructive empiricism.) The predicates in $\lambda U \mu$ are supposed to be identified not merely syntactically but in terms of the roles they play in the scientific language $L(\lambda U \mu)$. Sellars' dot-quotation operator can be taken to specify those roles. Treating predicates in this way guards against tricks performed by means of purely syntactic manipulations.

Now if $T(\lambda U \mu)$ establishes deductive logical relationships between some sentences containing only λ-predicates, then Craig's theorem on elimination shows that there is a (maximal) subtheory, say $T'(\lambda)$, of $T(\lambda U \mu)$ containing only λ-predicates which establishes the same logical relationships. The same remark applies to the Ramsey-sentence of $T(\lambda U \mu)$. (If, e.g., μ contains only P as its sole member, then $(E\pi)T(\lambda U\{P\})$ is the Ramsey-sentence of $T(\lambda U\{P\})$, where π is a variable replacing P throughout.)

Accordingly, theoretical terms are in a sense eliminable in the case of deductive logical systematization. We may put this more clearly as follows. Let us define:

(**IDS**) The predicates in μ of a scientific theory $T(\lambda U \mu)$ are **logically indispensable for deductive systematization** with respect to λ (and standard logical deduction) if and only if
(1) $T(\lambda U \mu)$ achieves deductive systematization between some sentences F and G in the language $L(\lambda)$;
(2) there is no (recursively axiomatizable) subtheory, say $T'(\lambda)$, of $T(\lambda U \mu)$ which achieves the same deductive systematization; nor is there a subtheory $T''(\lambda U \mu')$, $\mu' \subset \mu$, of $T(\lambda U \mu)$ which achieves the same systematization.

Thus we can grant that λ-predicates are always logically dispensable in the precise sense of (**IDS**).

   However, note that if $T(\lambda \cup \mu)$ establishes epistemic relationships between sentences containing only μ-predicates,
Craig's theorem has nothing immediate to say about that nor does
Ramsification really directly apply (even if the Ramsey-sentence
of a theory in a sense preserves its whole deductive structure).
But let us be charitable to an eliminationist and accept premise
(3), viz. that theoretical terms must have an observational
effect, so to speak.

   The second response that can be made against Craig's and
Ramsey's elimination methods is that they do not always preserve
the really central epistemic relationships in relation to the λ-
sentences. Especially they do not preserve all explanatory relationships and all inductive relationships. Let me now discuss
these matters briefly.

2. Recall the example about the solubility of gold and its
microexplanation. To quote Sellars, "microtheories not only explain why observational constructs obey inductive generalizations, they explain what, as far as the observational framework
is concerned, is a random component in their behavior, and in
the last analysis it is by doing the latter that microtheories
establish their character as indispensable elements of scientific explanation and as knowledge about what really exists"
(Sellars, 1963a, p. 122). In this - in my view correct and
defensible - objection to premise (4) of (**TDA**) the following two
claims can be discerned: (1) empirical generalizations require
(or at least may require) theoretical explanation, (2) the
explanation of at least some singular empirical statements requires theoretical explanatory concepts. Let us formalize the
empirical situation in a somewhat simplified manner in terms of
the following generalizations, assuming gold to have the solubility rate n_1 in at most q % of the cases and n_2 in the rest of
the cases:

(G1) $p(O_2/O_1) \leq q$
(G2) $p(O_3/O_1) > 1-q$
(G3) $(x)(O_2(x) \rightarrow \sim O_3(x))$.

Here $O_1(x)=x$ is a piece of gold, $O_2(x)=x$ is something that dissolves in aqua regia at rate n_1, $O_3(x)=x$ is something that dissolves in aqua regia at rate n_2, and p is a relative frequency-probability. The realist's claim now is that (G1) and (G2) are not lawlike generalizations (this is not a matter of their being probabilistic, by the way). They need the injection of scientific theory into them to make them lawlike.

A theory, S, designed to explain (G1) and (G2) is now given by

(A1) $p(T_1/O_1) = q$
(A2) $p(T_2/O_1) = 1-q$
(A3) $(x)(O_1(x)\&T_1(x) \rightarrow O_2(x))$
(A4) $(x)(O_1(x)\&T_2(x) \rightarrow O_3(x))$.

Here $T_1(x)=x$ is something with microstructure m_1 and $T_2(x)=x$ is something with microstructure m_2. T_1 and T_2 are theoretical predicates. (Note that I am using, for simplicity, the same variable x in the case of both the observational and theoretical predicates marking the same ontology for them. This simplification can be removed. For instance, (A3) would (or could) become $(x)(y)(O_1(x)$ & $T_1(y)$ & $R(x,y) \rightarrow O_2(x))$, where $R(x,y)$ specifies the relation between the observable item x and the (typically unobservable) item y.)

Now one may claim, as in effect does Sellars (1963a), p. 122, that (1) this theory S and hence its theoretical terms are needed to explain G_1 and G_2, viz. why O_1 is connected with O_2 with less than q % of the cases and with O_3 otherwise, and that (2) they are also needed to explain that something a which is O_1 also happens to be O_2. Thus (G1) is explained by deducing it from S, with dependence on its theoretical terms. That a which is O_1 is also O_2 is explained by S and the fact that a is T_1 rather than T_2. Clearly, $T_1(a)$ is essential to the explanation of this explanandum - or so it seems at least (a strict proof cannot be given a priori).

However, in his defense of instrumentalism Cornman claims that "if a system explains randomness by having the generalization that expresses the randomness as an epistemic theorem, then the system transcribed in either (i.e. Craigian or Ramsey-an) way also explains it" (Cornman, 1975, p. 177). But this is flatly wrong. For instance, in the case of deductive systematization explanation of course cannot be identified with deduction.

Consider e.g. the explanation of (**G1**). Cornman claims that C-**S**, the Craigian transcription of S, explains it because it has as its axiom the conjunction consisting of (**G1**) conjoined with itself a certain number of times. But that represents a clear case of self-explanation, which cannot of course qualify as a proper explanation.

I have discussed this particular problem in Tuomela (1973), chapter VII, where I also show that such an "explanation" does not satisfy the model of explanation (in the sense of so-called ε-arguments) presented in that book. In that same context I also discussed the explanation of empirical generalizations such as (**G1**) by means of the Ramsey sentence of a theory. I argued that the (deductive) explanation given in terms of the original theory is, in any case, better than the Ramseyan explanation already because in its case a stronger explanans is employed. In all, then, transcribed theories do not successfully compete with the original ones in explaining observational generalizations.

Let us next consider singular explanation. Is $T_1(a)$ after all essential to the explanation of why a, which is O_1, is also O_2? Let us first consider Ramseyan explanation. First we notice that $T_1(a)$ is an element of our explanans. We may therefore, at least for the sake of argument, grant that $\pi_1(a)$ be included as a conjunct under the scope of the existential quantifier quantifying over the predicate T_1 by means of a predicate variable π_1. (Then the lawlike observational generalizations, with theory injected into them, would be (**A3**) and (**A4**) changed by the use of π_1 and π_2 instead of T_1 and T_2 in them.) Given this much, one may claim, and Cornman does indeed claim, that the Ramsified explanans preserves the explanatory power of the original explanans (including $T_1(a)$ as its element). But here again (just as in the previous case) the Ramsified explanans is weaker in explanatory power than the original explanans. Furthermore, we must be able to use the variable π_1 replacing T_1 as a "quasi-constant" (e.g. in making reports such as $\pi_1(a)$). As I argued in Tuomela (1973), this practically amounts to realistic (as opposed to instrumentalistic) theorizing. And, anyway, if we think of the values of the variable π_1, the instrumentalist hardly will be able to show that they must be empirical predicates as his program requires.[1]

How about the Craigian case then? As the Craigian transcription of the original explanans gives us as the relevant empirical theorem only $O_1(a) \rightarrow O_2(a)$ conjoined with itself a num-

ber of times, but no law, our instrumentalist seems to be in
deep trouble. However, Cornman's remedy now is to use the generalization $(x)(x=a) \rightarrow (O_1(x) \rightarrow O_2(x))$ for the explanation. But this
won't do, because certainly this generalization cannot be a law
of nature (for its scope is a priori limited to one individual
only and it is thus equivalent to the singular statement $O_1(a) \rightarrow$
$O_2(a)$), and an explanation requires (implicitly or explicitly) a
law, I take it (see Tuomela, 1973, Chapter VII, and 1980b). Thus
the instrumentalist is not successful here either.

Let me conclude our discussion of the aqua regia -example
by emphasizing what I (contra e.g. Cornman) take to be central
about it. This example under my interpretation is based on the
assumption that the generalizations (**G1**) and (**G2**) are **not** in a
proper sense **lawlike**. That is why they require explanation, and
they can be explained only by means of the resources of the
scientific image and by in a sense injecting a theory into them.
The corresponding theory-laden generalizations (**A3**) and (**A4**) are
nomic and although they can of course be given deeper explanation they do not cry for it in the sense (**G1**) and (**G2**) do. The
fact that (**A3**) and (**A4**) are strict rather than probabilistic
generalizations is inessential. There are also other simplified
features in this example, e.g. that the observational predicates
O_1, O_2, and O_3 rather than their scientific counterparts occur
in the toy-theory S (cf. Section IV). In the general case it is
not warranted to think that the predicates of the manifest image
can be connected with the predicates of the scientific image by
means of correspondence rules (of the standard kind). Rather the
situation is such, in general, that only some kind of counterparts (in the scientific image) of the predicates of the manifest image can be connected with the predicates of the scientific image by means of some kind of correspondence rules (cf.
Chapter 9). Thus, in all, the aqua regia -example must be taken
with some salt.

We have above given reasons to think that theoretical concepts may be indispensable both for the (deductive) explanation
of empirical generalizations and singular empirical statements.
We may formulate this sense of indispensability more precisely
in terms of the following definition, where we use the relation
ε of deductive explanation of Tuomela (1973) and (1980b):

(**IDE**) The predicates in μ of a scientific theory $T(\lambda \cup \mu)$ are
logically indispensable for deductive explanation with

respect to λ and the explanatory relation ϵ if and only if

(1) for some (general or singular) sentence F in $L(\lambda)$, $T(\lambda \cup \mu)$ deductively explains F (in the sense of ϵ-explanation);

(2) there is no sentence $T'(\lambda)$ employing as its extra-logical predicates only those in λ which is entailed by $T(\lambda \cup \mu)$ and which also explains F (in the sense of ϵ-explanation); nor is there a subtheory $T''(\lambda \cup \mu')$ of $T(\lambda \cup \mu)$, with $\mu' \subset \mu$, with equal or better explanatory power.

Note that both the Craigian transcription of $T(\lambda \cup \mu)$ and its Ramsey sentence are theories (or sentences) of the type $T'(\lambda)$ spoken about in clause (2); explanatory power can be understood in the sense to be clarified in Chapter 9. We have above argued that some theoretical predicates are logically indispensable for deductive explanation in the sense of (**IDE**). But note that we have not shown that (2) holds for all observational theories, even if, e.g., our argument applies to all subtheories in $L(\lambda)$ of $T(\lambda \cup \mu)$. (Let us call this stronger statement (2') and the resulting definition (**IDE'**).) Furthermore, even if there were no such theory $T'(\lambda)$ in the sense of clause (2) of (**IDE**) or even in the sense (2') of (**IDE'**), there might be explanatory theories employing some explanatory observational predicates other than those of λ which would do the explanatory job. In order to be able to handle this, we may regard λ as representing the totality of the empirical predicates in the manifest image (to the extent this is a meaningful notion) to begin with. Another relevant point is that we have above in our above definition opted for the simplified **correspondence rule** approach instead of using the **reinterpretation** approach (alluded to in our above comments on the **aqua regia** -example). But, to keep matters simple, we shall below in our formulations rely on the correspondence rule account.[2]

What all this boils down to is that although we may have and do have tight and sound arguments for the indispensability of theoretical concepts in the sense of (**IDE**) in the case of some possible theories, we do not have a conclusive, logically tight proof in the case of the stronger (2'). I doubt that such proof can be found a priori for even fictitious theories, still less for, e.g., currently accepted scientific theories nor, what is most central, for the ultimate best-explaining ones.

3. In any case, our argumentation can be taken to support the falsity of the conclusions of (6) and (9) of (**TDA**) and in that sense to support realism both epistemically and ontologically. We can in fact summarize part of the argumentation by construct- ing a simple deductive argument for weak scientific realism. This argument will be stated in "pragmatic" terms to support the reasonableness of accepting weak scientific realism. The notion of indispensability for deductive explanation, viz. (**IDE'**), will play a key role here. Although I have not discussed my notion of deductive explanation in detail above, it should be emphasized that even a potential explanation requires a lawlike explanatory theory or law in its explanans (cf. Tuomela, 1973, Chapter VII, and 1980b). The argument for realism can be formulated as fol- lows, with λ representing the set of all observational (empiri- cal) predicates:

(**WR2**)(i)   Some scientific theoretical predicates are logically indispensable for factually true deductive explana- tion relative to λ (and ε).

(ii)   If some scientific theoretical predicates are logi- cally indispensable for factually true deductive ex- planation relative to λ (and ε), then it is reason- able to believe that these predicates refer to real (and typically unobservable) scientific entities.

(iii)   If it is reasonable to believe that some scientific theoretical predicates refer to real (and typically unobservable) scientific entities, then it is reason- able to believe that weak scientific realism is true.

(iv)   Therefore, it is reasonable to believe that weak scientific realism is true, viz. it is reasonable to accept it.

In premise (i) the notion of factually true explanation is used. It means explanation in which the premises of the explana- tory argument are factually true (cf. Chapter 9). Premise (i) can be supported by making reference to the above aqua regia - example and other similar examples. Also examples from actual scientific research may be cited (see Tuomela, 1973, Chapters VI and VII). In general, arguments to the effect that the framework

of the manifest image cannot sustain lawlike generalizations can
be used to support (i), for even potential explanations require
lawlike explanatory theories or laws (cf. the aqua regia -example). And even if there were lawlike generalizations within the
manifest image - contrary to our earlier arguments - some theoretical concepts might still be indispensable for deductive explanation for other reasons. In any case, we shall regard (i) as
true on the basis of our previous discussion (also see the
comments below).

Premise (ii) is also central. It connects actual or true
explanation with ontology in a certain way. (ii) can be supported by the general idea that true theories - ones we don't
presently seem to have - specify what there is in the world. (As
we shall argue in Chapter 9, a factually true theory must ideally also be the best-explaining theory. Thus we could have required in (ii), alternatively, that $T(\lambda \cup \mu)$ be a best-explaining
theory relative to λ-phenomena.) If now the observational
part of the world is unable at least in some cases to sustain
true explanations (in the sense of (i)) such explanations must
be based on the scientific image of the world and hence typically on the unobservable components of the world; and as we
continue to assume - what will be argued for in Chapter 7 - that
there are no ghosts, demons and the like, these unobservable
constituents must be scientific objects.

I am naturally assuming here that explanatory arguments
with true premises, viz. true explanations, especially causal
ones, do reflect the ontology of the world. Explanations reveal
the causal structure of the world, so to speak, and as I characterize what there is (and is not) in the world, in terms of
participation in causal interaction, we basically have what we
want here. Thus, if some theoretical predicate is indispensable
for true deductive explanation it is reasonable to regard it as
ontologically significant, too.

But wait a moment, we may expect Quineans and Putnameans
say at this point, for according to them truth and reference are
indeterminate. Therefore, there is no unambiguous way from factual truth to ontology. My brief answer here is that if truth is
taken to involve picturing and if the discussion is relativized
to a conceptual scheme or language no such indeterminacy follows, and this will be our strategy in this book. (See Sections
I and III of Chapter 6 as well as Section III of Chapter 9,
which also comments on the comparison of successor conceptual

schemes with their predecessors). Viewed in this way, (ii), will be non-contingently true, given our purported (partial) explication of factual truth in terms of picturing.

Note still two more things here. First, while theoretical predicates can be taken typically to refer to unobservable entities not all of them have to do so. As a matter of fact, the observable - unobservable dichotomy (which the empiricist badly needs) is not at all central here. But what is crucial is that all the entities that best-explaining (and factually true) theories posit qualify - be they in some empiricist or other sense observable or not.[3] Secondly, note that, while, for instance, the predicates 'means', 'is justified', and 'is morally good' may be regarded as indispensable for some purposes (though hardly for factual explanation), yet they should not (or at least need not) be taken to be ontologically significant. Thus indispensability and ontological significance go together only in the case of predicates used for directly describing the world, predicates that might be used in true scientific theories.

As to premise (iii), I regard it as true mainly in virtue of what we mean by weak scientific realism (and reasonable belief). In saying this I assume reference in (iii) to mean successful reference (rather than something like potential reference) and remind the reader that we are here dealing with Ersatz-reference if Sellarsian semantics is adopted (cf. Chapter 3). (WR2) is, however, meant to be open to different accounts of reference and ontology and should be read neutrally in this sense. Given all this, we have in (WR2) a logically valid and sound argument for weak scientific realism, I claim.

Let me still mention a closely related idea that can be used to construct a resembling argument for weak scientific realism. In this argument we start explicitly with the idea that the world (viz. any aspects and parts of it) is explainable scientifically (in a sense involving best-explaining scientific laws and theories; see Chapter 9) and proceed to claim that it is, however, the case that only the world as conceptualized in terms of the scientific image is capable of sustaining the required explanatory laws. Then the argument continues essentially following (WR2). Let me formulate this new argument as follows in more precise terms:

(**WR3**)(a) The world is scientifically explainable, viz. there are best-explaining theories about the world.

(b) If the world is scientifically explainable, then it is so explainable either when conceptualized in terms of the manifest image or when conceptualized in terms of the scientific image.

(c) The world is not thus scientifically explainable when conceptualized in terms of the manifest image (not even the observable aspects of the world are thus explainable).

(d) Therefore, the world (and even its observable parts) can only be scientifically explained when conceptualized in terms of the scientific image.

(e) If the world can only be scientifically explained when conceptualized in terms of the scientific image, then there are some best-explaining theories, which are conceptualized in terms of the scientific image and thus contain scientific theoretical predicates logically indispensable for the explanation of empirical phenomena.

(f) If some scientific theoretical predicates occur in some best-explaining theory (or theories), then it is reasonable to believe that these predicates refer to real (and typically unobservable) scientific entities.

(g) If it is reasonable to believe that some scientific theoretical predicates refer to real (and typically unobservable) scientific entities, then it is reasonable to believe that weak scientific realism is true.

(h) Therefore, it is reasonable to believe that weak scientific realism is true.

(**WR3**) is a logically valid argument. Is it sound? Premise (a) is one assumed by science and the scientific outlook in general and I shall not here specifically argue for it (in a sense this whole book constitutes a defense of this premise). Given that demons, spirits and the like can be excluded from the ontological furniture of the world - which we assume; cf. Chapters 7 and 10 - premise (b) can be regarded as acceptable. As to (c) let me just refer to our earlier discussions in this and the previous chapter (also cf. (i) of (**WR2**)). Premise (d) is a logical consequence of (a), (b), and (c). Premise (e) is true essentially in virtue of what best scientific explanation involves. See Chapter 9 for an account of best explanation, from which it follows that the involved theoretical predicates are indispensable for explanation.

Premise (f) basically restates the general **scientia mensura** -thesis defended in more detail in Chapter 7 and 10 (also see Chapter 9 for the equivalence of maximally informative factual truth and best explanation). Premise (g) is essentially the same as (iii) of (**WR2**). Note that while (**WR2**) explicitly relies on **true** theories (**WR3**) only speaks about **best-explaining** ones. However, if our argumentation in Chapter 9 below is sound, best-explaining and factually true theories will coincide. So, (**WR3**) has been argued to be both a valid and a sound argument for scientific realism.

Going back to the theoretician's dilemma argument (**TDA**), the above discussion has centered around the falsity of its premise (4). How about (5)? If epistemic relationships in (**TDA**) are taken to be just explanatory relationships I would not like to quarrel with (5), for, as will be seen, an (internal or epistemic) realist makes best-explaining theories the measure of reality. But if only logical deduction, for instance, is meant in (5), then of course it is unacceptable, too.

In his recent book van Fraassen (1980) claims that the above kind of "linguistic" approach (in terms of Craigian eliminability, etc.) to the problem of theoretical terms is completely misplaced: "no concept which is essentially language-dependent has any philosophical importance at all". I disagree with this claim (while agreeing that much philosophically uninteresting and spurious discussion can be found in the literature related to Craigian and Ramseyan elimination methods). My reasons for disagreeing are Kantian: unless one accepts the Myth of the Given one cannot strictly separate problems of language from problems related to the world, viz. there is no strict conceptual language-world distinction (recall Chapter 3 and Chapter 6 below). Model-theoretic or set-theoretic languages are no exception to this, even if in many cases they are more powerful than (simple) formal languages.

II. Theoretical Concepts Within Inductive Systematization

Up to now we have dealt with the role of theoretical concepts within deductive systematization and explanation only. It

is time to say something about inductive systematization and
explanation as well. The desirability and indispensability of
theoretical concepts within inductive contexts can be studied
much as we have above studied deductive ones. (In Niiniluoto and
Tuomela, 1973, this whole problem area has been systematically
discussed from various angles and I wish to refer the reader to
that work for details.)

Let us define two notions of inductive indispensability
relative to a notion of inducibility, I. These notions are
inductive analogues of (**IDS**) and (**IDE'**):

(**IIS**)   The predicates in μ of a scientific theory $T(\lambda\cup\mu)$ are
logically indispensable for inductive systematization
with respect to λ and I if and only if whenever $T(\lambda\cup\mu)$
establishes some inductive systematization with respect
to λ and I there is no theory in $L(\lambda)$ which achieves at
least the same inductive systematization; nor is there a
subtheory $T''(\lambda\cup\mu')$ of $T(\lambda\cup\mu)$, with $\mu'\subset\mu$, achieving at
least the same inductive systematization.

(**IIE**)   The predicates in μ of a scientific theory $T(\lambda\cup\mu)$ are
logically indispensable for inductive explanation with
respect to λ and I if and only if for some sentence G in
$L(\lambda)$, if $T(\lambda\cup\mu)$ inductively explains G (in the sense that
the inductive relation $\rho(G,T)$, as defined in Tuomela,
1981, obtains) and there is no theory in $L(\lambda)$, nor any
subtheory $T''(\lambda\cup\mu)$ of $T(\lambda\cup\mu)$, $\mu'\subset\mu$, which so inductively
explains G.

In (**IIS**) we mean by inductive systematization the following.

The theory $T(\lambda\cup\mu)$ achieves **inductive systematization** with
respect to λ and I if and only if for some statements K
and H of $L(\lambda)$
(a) (**T&K**)IH
(b) not KIH
(c) not (**T&K**) ⊢ H.

Here TIH is read "H is inducible from T", and inducibility may
be interpreted in several ways. For instance, inducibility may
mean positive probabilistic relevance: TIH if and only if
$p(H/T)>p(H)$.

Weaker notions of indispensability than those just defined can be obtained by requiring the definientia of (IIS) and (IIE) to hold only for all the subtheories in $L(\lambda)$ of $T(\lambda \cup \mu)$ rather than for all theories in $L(\lambda)$. In Niiniluoto and Tuomela (1973), Chapter 9, the logical indispensability of theoretical concepts within inductive systematization in a such a weaker sense is strictly shown for some (factually possible) monadic theories. Suitable theories without any non-tautological deductive consequences qualify as examples. Some of those examples also qualify to show the indispensability of theoretical concepts in the weakened version of (IIE), viz. for inductive explanation. However, it is here as hard as in the deductive case to strictly prove a priori that there are theoretical concepts indispensable in the unqualified senses of (IIS) and (IIE). These schemes, especially (IIE) can be used to support weak scientific realism completely parallelly with the case of deductively indispensable theoretical predicates. (I will not here explicitly write out this argument corresponding to (WR2).)

Within inductive contexts no elimination procedure corresponding to Craig's method in the deductive case can be defined, even if claims to this effect have been made in the literature. Thus Cornman (1975) has argued that the relation I of inducibility is quite comparable to the relation of deducibility in this respect. While it is not possible to go into the details here the fact of the matter is that Cornman is completely mistaken in his claim (see, e.g., Tuomela, 1978a, for arguments against Cornman).

Theoretical concepts can in many cases be shown to be **desirable** within inductive contexts. They may provide various gains for theoretization. In Niiniluoto and Tuomela (1973) such gains as measured by the degrees of corroboration, expected utility, explanatory power and lawlikeness are systematically studied in the context of conceptual and theoretical growth. Generally speaking, the idea is to compare the probabilities $p_0(G/E)$ and $p_1(G/E\&T)$ and to try to find cases in which $p_1(G/E\&T) > p_0(G/E)$, where p_0 is a probability measure for $L(\lambda)$ and p_1 for $L(\lambda \cup \mu)$, G and E are sentences in $L(\lambda)$ (or the sublanguage of $L(\lambda \cup \mu)$ confined to λ) and T in $L(\lambda \cup \mu)$. T can here be used as theoretical evidence, comparable to the observational evidence E as to its function. I shall not here go deeper into a discussion of the role of theoretical concepts within inductive systematization. (See Tuomela, 1981, for relevant remarks on the inductive-probabilistic explanation of singular statements.)

III. Quantificational Depth and the Methodological Usefulness of Theoretical Concepts

1. Besides those discussed in Sections I and II there are many
other arguments for showing the desirability or even the indispensability of theoretical concepts in science. Ultimately these
arguments seem to turn on the greater explanatory power of
theoretical concepts relative to observational ones; and this
supports scientific realism, as we have kept arguing. The considerations I have in mind are related to the gains due to the
use of theoretical concepts for the growth or increase of explained observational statements and of economy in theory-con-
struction, for induction and for many other tasks and aims,
e.g., increase in observational richness due to non-eliminable
theoretical terms in the so-called model-theoretic Ramseyan
sense. Let me below take up some such considerations which I
have discussed at length in Tuomela (1973).

Suppose that we are given a first-order scientific theory
axiomatized by the sentence $T(\lambda \cup \mu)$, with $\lambda \cap \mu = \emptyset$. As before, λ-
predicates may be regarded as "observational" or "non-theoreti-
cal" or "old" with respect to the theory $T(\lambda \cup \mu)$ while those in μ
are "theoretical" or "new" with respect to it. This theory can
be transformed into its distributive normal form (in the sense
of Hintikka). Now as a theory can be regarded as a logically
closed general sentence, it will contain nested quantifiers. Let
us call the **depth** of T the number of layers of quantifiers it
contains at its deepest. The depth of T is (with certain minor
qualifications) equal to the number of individuals (conceptual
individuals in the logical sense) considered in their relation
to each other in T (see Hintikka, 1973a, Hintikka and Tuomela,
1970, and Tuomela, 1973, for more details). Let the depth of T
be **d**. We denote this by $T^{(d)}$. Now $T^{(d)}$ has a logically equivalent expansion (a disjunction of constituents in the sense of
Hintikka) at some greater depth d+e. Thus $T^{(d)} \leftrightarrow T^{(d+1)} \leftrightarrow, \ldots, \leftrightarrow$
$T^{(d+e)} \leftrightarrow \ldots$, ad infinitum, where $\leftrightarrow$ represents logical equivalence.

We may now ask what $T^{(d)}(\lambda \cup \mu)$ says exclusively about the
members of λ, viz. λ-predicates and λ-discourse. One answer to
this question is given by the set of deductive consequences of
$T^{(d)}(\lambda \cup \mu)$ which contain no members of μ. But this is not the
only relevant answer. For we may define a relevant notion of the

reduct of $T^{(d)}(\lambda \cup \mu)$ into the vocabulary λ. This reduct operation was defined in exact terms by Hintikka and Tuomela (1970) and I will not here go into the technical details of this matter. Let us assume the availability of this depth-dependent reduct operation and call it r_λ.

A counterpart of Craig's so-called interpolation theorem can be proved in the theory of distributive normal forms of Hintikka. Applied to the present case it gives the following two-layered diagram:

$$T^{(c)} \leftrightarrow T^{(c+1)} \leftrightarrow \ldots \leftrightarrow T^{(d)} \leftrightarrow \ldots \leftrightarrow T^{(d+e)} \leftrightarrow \ldots$$

$$\downarrow \qquad\qquad \downarrow \qquad\qquad \downarrow \qquad\qquad \downarrow$$

$$r_\lambda(T^{(c)}) \leftarrow r_\lambda(T^{(c+1)}) \leftarrow \ldots \leftarrow r_\lambda(T^{(d)}) \leftarrow \ldots \leftarrow r_\lambda(T^{(d+e)}) \leftarrow \ldots$$

Diagram 5.1.

Here single arrows stand for logical implication and double ones for logical equivalence. The interesting thing about the diagram is that in the lower row we have only single arrows. Thus, e.g., while $r_\lambda(T^{(c+1)})$ logically implies $r_\lambda(T^{(c)})$ the converse is not the case.

While the above diagram is in many ways idealized it is not without methodological interest. Let us thus consider what we may call theoretical generalization (or theoretization). Suppose we have empirically come upon a generalization S such that the distributive normal form of S, say $S^{(d)}$ is just $r_\lambda(T^{(d)})$ such that S cannot be given a distributive normal form of depth less than d. Then we ponder upon the explanation of $S^{(d)}$ and invent the theory $T^{(d)}(\lambda \cup \mu)$, which, we assume, not only entails $S^{(d)}$ but gives a deductive ε-explanation of it (in the sense of Tuomela, 1973, and 1980b). But given $T^{(d)}$ we may move towards logically equivalent expansions $T^{(d+e)}$, $e > 0$, of it. Now $r_\lambda(T^{(d+e)})$ logically entails $S^{(d)}$ but is not so entailed by it. This means a deductive gain in observational richness.

Let us assume that we are able to adequately measure the informational content (cont) of our sentences (see, e.g., Tuomela, 1973, and Niiniluoto and Tuomela, 1973 on this). Then

(1) $\lim_{e \to \infty} \text{cont}(r_\lambda(T^{(d+e)})) - \text{cont}(r_\lambda(T^{(d)}))$

measures the amount of the **deductive gain in observational richness** due to the introduction of the theoretical concepts in μ in
this situation.

There is also another gain due to theoretical concepts that
our diagram illustrates. Suppose there is no shallower sentence
logically equivalent to S than $S^{(d)}(= r(T^{(d)}))$. Now if we, however, want to describe the world in terms of shallower sentences
and thus in more economical ways than $S^{(d)}$ we may do it in terms
of $T^{(c)}$, for c<d. (We assume here that we have invented a theory
$T^{(d)}$ which indeed has shallower equivalents.) Then, considered
from the point of view of λ-sentences, we here get a **deductive
gain in descriptive economy** due to using theoretical concepts
(relative to $S^{(d)}$). This gain can be measured by

(2) $\text{cont}(r_\lambda(T^{(d)})) - \text{cont}(r_\lambda(T^{(c)}))$.

The above possible gains due to theoretical concepts indicate of course that theoretical concepts can be methodologically
desirable and are so when either (1) or (2) is positive. If and
when the single arrow between the two rows of Diagram 5.1.
represents not only logical implication but deductive explanation (in the sense of the explanatory relation ε) cases of
indispensable theoretical concepts in the standard sense (**IDE**) -
and perhaps also (**IDE'**) - can be found (cf. our discussion of
the aqua regia -example). But Diagram 5.1. does not help to make
any theoretical concepts logically indispensable in the sense
(**IDS**), for it is always equally logically possible to "invent" the sentences stronger (and deeper) than $S^{(d)}$ on the
lower row of the diagram as it was to invent $T^{(d)}$.

2. Let me now illustrate more concretely the usefulness of
theoretical or "new" concepts by means of two examples (previously discussed in Tuomela, 1973, pp. 158 - 162). First, we
consider an artificial example best described by reference to
elementary arithmetic. Suppose that we are in a position to make
observations concerning a single monadic property P of individ-

uals in an ordering. The ordering is given by a two-place relation R to be interpreted as immediate succession. The axioms
governing R are given in a first-order language with identity as
follows:

(A1) $(x)(Ey)R(x,y)$;
(A2) $(x)(y)(R(x,y) \rightarrow -R(y,x))$;
(A3) $(x)(y)(z)(R(x,y)\&R(x,z) \rightarrow y=z)$;
(A4) $(x)(y)(z)(R(y,x)\&R(z,x) \rightarrow z=y)$;
(A5) $(Ex)(y)(-R(y,x))$;
(A6) $(x)(z)(-(Ey)R(y,x)\&-(Ey)R(y,z) \rightarrow x=z)$.

These axioms describe an immediate succession-ordering with a
unique first element but no last element. The immediate and
distant (finite) successors of the first element can be correlated with the natural numbers 2,3,4,...
    Let us assume that a theorist can (directly or indirectly)
"observe" which individuals have the property P. Assume that,
starting from the beginning, individuals no. 2,3,5,7,11, and 13
are found to have this property out of the 13 first individuals.
How can one try to "systematize" or to "explain" these observations? How can one obtain predictions concerning new individuals? It is intuitively obvious what the tempting suggestion
here is, viz. that the predicate P belongs to those and only
those distant and immediate successors of the first individual
which are correlated with prime numbers. It is also clear that
this conjecture cannot be expressed in finite terms solely by
means of P and the successor relation. Hence we have here a
clear-cut (though artificial) example of the need to resort to
"theoretical" (auxiliary) concepts.
    One thing our imaginary theoretician could do here is to
expand our vocabulary so as to be able to formulate a fragment
of elementary number theory. To our "observational" theory (formulated in the vocabulary $\lambda=\{P,R\}$) he adds a set μ consisting of
two new three-place predicates A and M representing the addition
and multiplication operations, and a few new axioms to obtain a
richer theory $T(\{P,R\}\cup\{A,M\})$. Four of these axioms will state
that A and M are functions defined everywhere, and the others
can be modified versions of the usual recursion equations for
addition and multiplication (let us call them **(A7 - A14)**) plus
the following:

(A15) $(x)(P(x)\leftrightarrow(y)(z)(y\cdot z=x\to y=1\vee z=1)\&x\neq1)$.

   Here we have, of course, used dot as a shorthand which can
be eliminated in favor of the relation M. By means of axiom
(A15) we predict for any given future individual in our "observational" sequence whether it has P or not.

   Our richer theory $T(\{P,R\}\cup\{A,M\})$ with its fifteen axioms
(the deepest has depth 5) has thus been specified clearly
enough. Here we will not undertake the laborious task of writing
it out in its distributive normal form, nor is this needed for
our purposes. Neither can the axioms of the "observational"
theory axiomatized by the successive reducts

$$r(T^{(d)}(\{P,R\}\cup\{A,M\}))\qquad(d=5,6,7,8,\ldots)$$

into the language $L(\{P,R\})$ be easily transformed into their distributive normal forms. However, we can see what their essential
features would be like. In order to distinguish individual no. d
from the others (so as to be able to say whether it has the
predicate P or not) by means of the successor relation alone we
need d+1 layers of quantifiers. Thus an essential part of what
$r(T^{(d)}(\{P,R\}\cup\{A,M\}))$ says is to list which individuals among
those numbered $1,2,\ldots,d-1$ have P and which ones do not. In
brief, the poorer theory $r(T^{(d)}(\{P,R\}\cup\{A,M\}))$ $(d=5,6,7,\ldots)$ specifies which individuals have P, by enumerating them one by one,
while the richer theory presents us with a definite law governing the distribution of P.

   It is striking how the introduction of "theoretical" (auxiliary) concepts can give us some real gain. The richer theory
has deductive consequences concerning P and R beyond those of
the reducts of any of its successive expansions. If one starts
from any of these as one's poorer theory, a gain of the first
kind is obtained. For instance, in the "experimental" situation
described in our fictitious example, a natural starting-point
would have been $r(T^{(14)}(\{P,R\}\cup\{A,M\}))$. Furthermore, starting
from any reduct with depth >5, a gain in economy (gain of the
second kind) is obtained also, in that the richer theory can be
axiomatized with fewer layers of quantifiers. It seems to us
that the feeling of much greater "insight" into and "appreci-
ation" of the situation that the richer theory seems to give us
here is partly a reflection of these clearly definable features
of the underlying logical situation.

Although the confirmation theory of relational (polyadic)
generalizations is notoriously underdeveloped, it is clear that
the poorer theory (lower row) in which we present our general
hypothesis, so to speak, by enumerating the different cases
cannot in any case be highly confirmed or corroborated by finite
evidence, for such evidence cannot distinguish between this theory and a great many competing theories. In contrast, any adequate confirmation theory presumably ought to show how the
richer theory is confirmed by its instances (cf. inference to
best explanation). Thus the gain obtained by enriching the conceptual basis may in the last analysis turn out to be as closely
connected with the inductive properties of theories as with
their purely deductive ones.

Recall that it is possible to axiomatize some standard
version of set theory in first-order logic. As a great part of
mathematics can be developed within set theory we can thus in
principle formalize a great many actual scientific theories,
even if the task may seem quite hopeless to carry out in practice (see Field, 1980, for illustrations). (However, in some
interesting cases infinitary first-order languages seem better
suited for such formalizations - and our above results do not
directly apply to such languages.)

An example from real science is offered to us by certain
sociometric investigations (viz. attempts to explain the resulting sociograms). In these investigations contacts between individuals (a two-place relation) are recorded to form a sociogram,
together with certain further properties of these individuals.
Then new concepts incorporated in certain new axioms are introduced to explain these observations rather in the manner sug-
gested above, so that our gains in observational richness and
descriptive economy are obtainable. However, in this connection,
we cannot undertake a detailed methodological investigation of
these sociometric studies.

As our second example we shall discuss a gain in descriptive power within a version of Ohm's theory circuits. We shall
below mostly rely on Simon's axiomatization (Simon, 1970), but
some reformulation in translating the theory into our framework
is needed. (Here we cannot even attempt to give a full and rig-
orous first-order standard formalization of the theory.)

We consider a theory $T(\lambda \cup \mu)$ where $\lambda=\{r,c\}$ and $\mu=\{v,b\}$. Here
the observational terms r and c represent, respectively, the
resistance and current intensity of the components of any system

or domain D to which the theory is applied. They are (mathematically) interpreted as non-negative real valued functions defined on such a D. The theoretical terms v and b represent,
respectively, the voltage and the internal resistance of a
system S. They are assumed to be mathematically interpreted as
real constants. Our small theory $T(\lambda U\mu)$ of circuits is assumed
to satisfy the following axiom for every element in its domain

(A) $c = v/(b+r)$.

Let us now consider the definability of voltage and internal
resistance in terms of resistance and current intensity within
T. The definability of our theoretical concepts somewhat depends
upon the cardinality of the domains D into which the theory is
applied. If D contains one single element, v and b clearly
remain "underidentified" as **(A)** now implies

(A') $v = (b+r_1)c_1$

where r_1 and c_1 respectively designate the values of the resistance and current intensity of this single component. If D contains two elements with distinct r's and c's, then **(A)** gives us
explicit definitions (and unique identifiability) for v and b in
terms of r and c as follows:

(A") $b = (c_2r_2-c_1r_1)/(c_1-c_2)$; $v = c_1c_2(r_2-r_1)/(c_1-c_2)$

But in the general case when D contains more than two elements,
we obtain explicit definitions for v and b:

(L) $c_j=c_kc_1(r_1-r_k)/(c_1r_1-c_kr_k+(c_k-c_1)r_j)$

must hold. Notice that **(L)** is an observational law statement.
Let us denote the requirement (statement) that D contain at
least two elements with distinct c's and r's by **(B)**.

   Now consider the extended or full Ohmic theory $T'(\lambda U\mu)$
axiomatized by the conjunction of **(A)** and **(B)**. In this extended
theory the theoreticals v and b are eliminable by explicit definitions (of the form **(A")**) in terms of c and r. However, the
observational subtheory **(L)** obtained from the full theory
$T'(\lambda U\mu)$ is in a clear sense theoretically complex and uneconomi-
cal relative to the full theory where b and v are present. For

within T'($\lambda U \mu$) we can, for those purely theoretical purposes
where we operate solely or primarily with b and v, get along
with the formula (**A**), whose quantificational depth is 5 (in a
standard first-order formalization).[4] But if we eliminate b and
v from the full theory T'($\lambda U \mu$) then we for all purposes have to
go back to the complex formula (**L**), whose depth is 15.

In other words, for some theoretical purposes at least, we
obtain a definite economy gain in depth (of the magnitude 15-5=
10, if you like) due to the employment of the theoretical concepts v and b in this Ohmic theory T'($\lambda U \mu$).

IV. A Scientific Realist's View of the Role of
Theoretical Concepts

1. Let us return to the ways of thinking that guided our construction of Diagram 5.1. We can in fact expand the diagram to
represent ideas which are even more suited to realism than those
discussed before. For we may in a sense invert the previous
picture and start with the assumption that we are given a "core"
theory H, which is solely in the vocabulary μ. After all, this
is how at least the "hard" sciences can be thought to operate.
Scientists often invent theories employing only highly theoretical terms and describe their subject matter of interest in terms
of them (see, e.g., Bunge, 1967, for illustrations of axiomatized physical theories without any empirical content). However, every scientific theory must be empirically testable - at
least indirectly, via auxiliary theories and assumptions. It can
also analogously be required to be of use in explaining empirical phenomena (cf. premise (3) of (**TDA**)).

Let us assume that H has been rendered in its distributive
normal form at depth d. This sentence H(d) we assume to be
identical with $r_\lambda(T(\lambda U \mu))$, viz. with the reduct of T($\lambda U \mu$) into
the vocabulary λ. We may thus expand our two-layered Diagram
5.1. into the following three-layered diagram (cf. Hintikka,
1973b, Tuomela, 1973):

$$r_\mu(T^{(c)}(\lambda \cup \mu)) \;\leftarrow\ldots\leftarrow r_\mu(T^{(d-1)}(\lambda \cup \mu)) \leftarrow r_\mu(T^{(d)}(\lambda \cup \mu)) \leftarrow$$
$$\uparrow \qquad\qquad\qquad\qquad \uparrow \qquad\qquad\qquad \uparrow$$
$$T^{(c)}(\lambda \cup \mu) \quad\leftrightarrow\ldots\leftrightarrow\quad T^{(d-1)}(\lambda \cup \mu) \quad\leftrightarrow\quad T^{(d)}(\lambda \cup \mu) \quad\leftrightarrow$$
$$\downarrow \qquad\qquad\qquad\qquad \downarrow \qquad\qquad\qquad \downarrow$$
$$r_\lambda(T^{(c)}(\lambda \cup \mu)) \;\leftarrow\ldots\leftarrow r_\lambda(T^{(d-1)}(\lambda \cup \mu)) \leftarrow r_\lambda(T^{(d)}(\lambda \cup \mu)) \leftarrow$$

$$r_\mu(T^{(d+1)}(\lambda \cup \mu)) \;\leftarrow\ldots\leftarrow r_\mu(T^{(d+e)}(\lambda \cup \mu)) \leftarrow\ldots$$
$$\uparrow \qquad\qquad\qquad\qquad \uparrow$$
$$T^{(d+1)}(\lambda \cup \mu) \quad\leftrightarrow\ldots\leftrightarrow\quad T^{(d+e)}(\lambda \cup \mu) \quad\leftrightarrow\ldots$$
$$\downarrow \qquad\qquad\qquad\qquad \downarrow$$
$$r_\lambda(T^{(d+1)}(\lambda \cup \mu)) \;\leftarrow\ldots\leftarrow r_\lambda(T^{(d+e)}(\lambda \cup \mu)) \leftarrow\ldots$$

Diagram 5.2.

Let us now think that we use the theory $H^{(d)}$ to explain the empirical law $S^{(d)}$ $(= r_\lambda(T^{(d)}(\lambda \cup \mu)))$. In order to be able to explain S by means of H these two general sentences must be appropriately connected. Let us here first think that we can do the explaining deductively and that a suitable "correspondence rule" or "interpretative system" can be used for the task.[5] Indeed, the theory $T^{(d)}(\lambda \cup \mu)$ itself qualifies as such an interpretative system. Its content over and above the content of H is

(3) $\operatorname{cont}(T^{(d)}(\lambda \cup \mu)) - \operatorname{cont}(H)$

while its content over and above S is

(4) $\operatorname{cont}(T^{(d)}(\lambda \cup \mu)) - \operatorname{cont}(S)$.

It is also worth noting that the expression

(5) $\operatorname{cont}(T^{(d)}(\lambda \cup \mu) - \lim\limits_{e \to \infty} \operatorname{cont}(r_\lambda(T^{(d+e)}(\lambda \cup \mu)))$

measures the extent to which theoretical concepts help us to say something informative about the world (viz. especially the unobservable aspects of the world) over and above what the observational concepts enable us to say.

Let us now consider the following key expressions (cf. Hintikka, 1973b):

(6) $\lim_{e \to \infty} \text{cont}(r_\lambda(T^{(d+e)}(\lambda U \mu)))-\text{cont}(S)$

(7) $\lim_{e \to \infty} \text{cont}(r_\mu(T^{(d+e)}(\lambda U \mu)))-\text{cont}(H)$

Now it is surely desirable that the limit expression (5) be as great as possible - the explanans should have many more observational consequences than only the explanandum. (6) just measures such excess observational strength of the explanans. But (7) should be small, at least typically. Why? Suppose H is a mature theory. Then the interpretative system should not add to its content, to what it says about the members of μ. For that would represent "theoretical adhockery" - inventing new theoretical axioms and assumptions to enable the core theory to perform its empirical explanation task. Ideally we may want to require (7) to equal zero.

Under what circumstances can (7) be equal to zero? This will happen when the maximal subtheory of $T^{(d)}(\lambda U \mu)$ in the language $L(\lambda)$ is finitely axiomatizable at a depth not greater than d. A sufficient condition for this is the definability of the members of λ in terms of those of μ within $T^{(d)}(\lambda U \mu)$ by means of noncreative definitions of depth less than or equal to d. For instance, explicit definitions of the members of λ in terms of those of μ (relative to $T^{(d)}(\lambda U \mu)$) would be such definitions.

The present methodological requirement of blocking theoretical adhockery, although formulated within the correspondence rule approach, reflects the general realistic idea of giving the common sense world or the manifest image of the world a reinterpretation in terms of the scientific image - the world as described by means of the best-explaining scientific theories (cf. above Chapter 2 and Sellars, 1963a, pp. 125 - 126). This idea of redefining observational terms is in fact compatible with what we said in Chapter 2 about the relationship between the manifest image and the scientific image. Our redefining correspondence rules reflect the "seeming-being" dialectic between the two images. They can be viewed syntactically and thus they need not be taken to identify any more the senses or the denotations of the observational terms with the conglomerations of the theoretical terms in the definientia. A strong realist

can   go even further and regard them as statements to the effect
that the objects of the manifest image (viz.   the   observational
framework) do not really exist - that there really are no   such
things.   Thus within strong realism both the senses (or   intensions)   and   denotations of observational terms   are   abandoned.
Weak realism, we recall, does not regard the manifest image in a
priori conflict with the scientific image and accordingly grants
that   the   entities of the manifest image may exist.   But it   is
open   to weak realism to say that they do not **really** exist   (and
to   construe **"real"** reference and denotation   accordingly)   - if
the   force   of   'really' is cashed out in terms   of   explanatory
power.   On   the   pragmatic level both weak   and   strong   realism
require   that in principle it should be possible that the theoretical   terms take over the reporting role of the   observational
ones.

      All this is of course quite the opposite to what empiricism
requires.   To   recall an extreme position,   at one time   logical
empiricists   required that all theoretical terms   be   explicitly
defined in terms of observational ones.   (Cf. Russell's "supreme
maxim" of philosophizing: "Whenever possible, logical constructions   are   to be substituted for inferred   entities";   Russell,
1917, p. 150.)

      A Kantian remark relevant to the internal realism advocated
in   this   book may be made at this point.   It is that   what   the
theory $T(\lambda \cup \mu)$ says about the world is strongly dependent on   its
conceptual   system.   This   can   be seen in two   different   ways.
First,   the predicates of λ are interwoven with those of μ (and
vice   versa).   Using   our earlier expressions, we   may   clearly
measure this in terms of the expression.

(8) $\mathrm{cont}(T^{(d)}(\lambda \cup \mu)) - \lim_{e \to \infty} \mathrm{cont}(r_{\lambda}(T^{(d+e)}(\lambda \cup \mu)) \& r_{\mu}(T^{(d+e)}(\lambda \cup \mu)))$.

In   general (8) is positive.   It becomes zero when   the   maximal
observational   and theoretical subtheories are   content-indepen-
dent (which may mean several things depending on how the measure
cont   is   defined).   Another and perhaps still deeper   sense   of
interwovenness is the following.   We may distinguish between the
depth   content   and the surface content of a theory rendered   in
its distributive normal form (see   Hintikka,   1970, 1973a).   The
surface   content   is obtained on the basis of the standard   distributive   normal form of the theory including logically   incon-

sistent constituents. When content is defined for distributive normal forms from which logically inconsistent constituents have been eliminated, we are dealing with depth content. The Kantian point now is that while the depth content of a sentence can be taken to reflect or picture the world in a straightforward sense - to be discussed below in Chapter 9 - there is no recursive method for eliminating inconsistent constituents and for arriving at depth contents, for that would require first-order predicate logic to be decidable (which we know it isn't).

2. Let us finally try to remove some of the idealizations involved in our earlier diagrams in order to get to an adequate scientific realist's view of theorizing. To begin with, scientific experimentation and data gathering typically use concrete "operationalized" terms, viz. terms related to measurement devices. Let us call such terms (measurement instrument readings, etc.) experimental terms of λ''-terms. Data are basically collections of singular λ''-statements.

The next step is to formulate empirical generalizations on the basis of such evidence. They are couched in the standard language of the manifest image and involve idealized terminology (relative to experimental terminology). Thus when measuring, say, electric current intensity, degree of solubility or degree of extrovertness, to mention quite different examples, a scientist clearly distinguishes between these empirical terms from the terms he uses when measuring what they express. Let us call these (possibly idealized) empirical terms λ'-terms.

Now the question may be raised whether our earlier set λ of observational predicates should be taken to be just λ'. The answer is that indeed it can in many contexts and that has in fact been the favored interpretation in our above discussion. But we have also noted that while this suits the correspondence rule approach it does not suffice for the realist's reinterpretation account. Indeed, we already found it necessary to speak of counterpart predicates and to introduce their set λ^*. Let me now make the suggestion that, in order to be able to use the notation of our earlier diagrams, we below take λ to be not λ' but λ^*, viz. the set of counterpart predicates. This is a notational suggestion of no importance in itself, but behind it lies an important philosophical point which I shall make by reference to an example.

Let us consider temperature. We can feel that iron is hot, water is lukewarm and that air is cool, for example. This phenomenological notion of temperature is the everyday, common sense notion of temperature. We have various beliefs related to it. In the case of gases we know that they approximately obey the Boyle-Mariotte law relating temperature and pressure. However, the kinetic theory of gases deals with different notions of temperature and pressure, which we may call the successor concepts or counterpart concepts within the scientific image of the phenomenological concepts, which can be regarded as concepts within the manifest image. Accordingly it is not the phenomenological notion of gas temperature (represented by the predicate 'temperature$_m$') but rather the scientific notion (represented by 'temperature$_s$'.) that is identical with the average kinetic energy of gas molecules. (In fact the phenomenological notion of temperature has several counterpart notions in the scientific image, depending on domain. Thus while temperature in a gas is mean kinetic energy, it is something different in a solid body and again different in a plasma.)

We may thus assume that the compound predicate 'average kinetic energy' belongs to μ and claim that temperature$_s$ = average kinetic energy, where a λ-predicate is explicitly defined in terms of μ-predicates within the kinetic theory of gases (= $T(\lambda U\mu)$). As said, temperature$_s$ is the counterpart of temperature$_m$, which is a notion of the manifest image. In our earlier terminology, 'temperature$_m$' is in λ', 'temperature$_s$' in $\lambda^{\star}$, and we presently identify λ with $\lambda^{\star}$.

When speaking of claims (or theories) rather than concepts we can also say that the theory $T(\lambda U\mu)$ correctively explains empirical generalizations, say $S'(\lambda')$, by deductively explaining general sentences $S(\lambda)$, given that $S(\lambda)$ is a counterpart claim within the scientific image of the empirical generalization $S'(\lambda')$; cf. our definition of corrective explanation to be given in Chapter 9. Note that we cannot have pure deductive explanations of this sort between two conceptual systems or languages such as the manifest image and the scientific image, for that would make the manifest image autonomous with respect to the scientific image (cf. note 1). But we can have corrective explanation in the illustrated sense.

Now we may summarize the whole situation by means of the following five-layered diagram, which makes Diagram 5.2. less idealized. To save space, I make explicit only one level of

depth and rotate the diagram 90 degrees to the right into another spatial position (so that the depths d grow when moving
downwards):

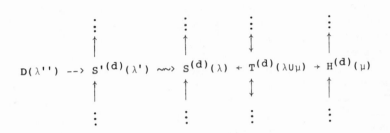

Diagram 5.3.

Here --> means inductive generalization on the basis of the
singular data $D(\lambda")$ yielding the empirical generalization
$S'^{(d)}(\lambda')$ as a result. We need not always get a neat implication
between the sentences $S'^{(d+e)}(\lambda')$ and $S'^{(d+e-1)}(\lambda')$, e>0, but
let us here idealizingly assume so. ~~> means reinterpretation
or translation of λ' into λ with the consequence that $S^{(d)}(\lambda)$
will be the counterpart claim within the scientific image of the
empirical generalization $S'^{(d)}(\lambda')$. $S'^{(d)}(\lambda')$ is typically only
an approximation of $S^{(d)}(\lambda)$ and the latter corrects the former.
The core theory $H^{(d)}(\mu)$ as "interpreted" by $T^{(d)}(\lambda \cup \mu)$ deductively explains $S^{(d)}(\lambda)$ and, intermediately, correctively explains
$S'^{(d)}(\lambda')$ (cf. Ch. 9 on corrective explanation and translation).

I do not want to claim that Diagram 5.3. is fully realistic. In fact it is not. (For one thing, it is confined to
deductive explanation only.) But I do claim that it represents
conceptually interesting features of scientific theorizing,
viewed from a scientific realist's point of view.

Note that one can of course investigate the desirability
and indispensability of theoretical predicates also within corrective explanation and within the set-up of Diagram 5.3. Mutatis mutandis, one can then write up arguments for (weak) scientific realism to match our earlier (**WR2**) and (**WR3**). (I will
spare the reader the details, but basically we just use the
phrase 'corrective deductive explanation' in the place of 'de-

ductive   explanation'   when defining indispensability   and   best
explanation.)

The previous and this chapter have,   I venture,   shown that
scientific realism is science's own philosophy, if there is one.
Scientific   realism   is   a philosophical   doctrine   which   takes
science   more   seriously than any other philosophical   view.   We
have reason to propose to scientists endorsing   epistemological
or   ontological   empiricism   in their words or   deeds   (such   as
propounders of an operationalist interpretation of quantum mechanics   and   many   psychologists and sociologists) that   they   in
stead should take a serious look at scientific realism.   Even if
a   working scientist does not all that often face   such   "fundamental" questions as we have discussed here,   the **methodological**
attitude   of realism towards actual scientific research   differs
from the attitude had by strict empiricists and positivists (cf.
also   Chapter 10).   Scientific realism is a clear alternative to
positivism   and   - if   the arguments of the   previous   and   this
chapter carry any weight - a superior philosophical view on   the
whole.

CHAPTER 6

INTERNAL REALISM

I. Metaphysical and Internal Realism

1. Hilary Putnam has stressed, in his recent writings, the distinction between metaphysical and internal realism (see e.g.
Putnam, 1978, 1981, 1982a, and 1982b). In what follows I shall
examine issues related to this distinction, or to something
related to this distinction. We should note, to begin with, that
the distinction is in fact a Kantian one. Kant came in his philosophy to oppose "transcendental realism" and to endorse "empirical realism" (and the "transcendental idealism" associated
with it). Putnam's metaphysical realism corresponds here,
roughly, to transcendental realism, and his internal realism to
Kant's empirical realism (cf. Chapter 1).

According to Putnam (1978) (scientific) realism is a kind
of an explanatory theory which comprises the following principles:
(1)   The terms of mature scientific theories typically refer.
(2)   The laws of mature scientific theories are typically
        approximately true.

This theory is assumed to explain scientists' behavior, if we
presume, in addition, that they believe that the hypotheses (1)
and (2) are true (of their fields of research). In Putnam's view
we can derive from these facts a workable explanation because
(1) and (2) are true (i.e. realism is a true doctrine). Consequently, according to him realism explains both the success of
science and the progress of science towards ever more truthlike
theories. Moreover, claims Putnam, such a brand of realism can
give a general explanation of why language use helps language
users to achieve their goals: speakers, namely, reflect the
world - their environment - by forming symbolic representations
of it (these representations have to do with correspondence
between expressions and sets of real objects in the sense of a
Tarskian theory of truth).

     As we noted already in Chapter 4 (note 4) against Putnam,
realism of the above kind can work only if it can be thought to
rely on a realist interpretation of truth as correspondence. As
such the hypotheses (1) and (2) are neutral with respect to both
the various interpretations of the concept of truth and the
various versions of realism. But the distinction metaphysical -
internal realism has to do, more than with anything else, with
different interpretations of the concept of truth.
     Metaphysical realism is a view according to which (a) the
world is "ready-made": it consists of a fixed set of objects
which are independent of human mind (cf. Chapter 3, Sections
I.1. and II.2.). Furthermore it nolds that (b) there is one and
only one true and complete description of the world. Here (c)
truth is a radically non-epistemic notion, i.e. correspondence
between linguistic expressions and the world. (Putnam, 1981, p.
49, labels this also the externalist perspective.)
     Internal realism (or the internalist perspective) on the
other hand denies all of the claims (a), (b) and (c). It is
characteristic of this view to hold that (a') the question "What
objects does the world consist of?" only makes sense within a
theory or description. Accordingly, the world is in a sense
"man-made" or "processed" through a human conceptual scheme (cf.
Section II). Typical versions of internalism also hold that (b')
the world can be described in several true and complete but in
some sense rival ways. Furthermore, according to it (c') truth
is an epistemic (and theory-dependent) notion.

2. Putnam has attempted to demonstrate the faults of metaphysical realism in two ways. First, he has presented model-
theoretic logical arguments against the non-epistemic notion of
truth. Unfortunately these arguments have relied on very questionable, if not incorrect, semantic presuppositions (for criticisms, see Koethe, 1979, Tuomela, 1979, 1980a, Pearce and Rantala, 1982a, 1982b, Lewis, 1984). I shall not here discuss them in
detail; in fact I will only briefly comment on the central argument of Putnam's recent book (Putnam, 1981).
     Putnam's most direct attack against metaphysical realism is
perhaps to be found in his **Reason, Truth and History** (Putnam,
1981). There he argues, roughly, that if meanings (including
references) "ain't in the head" or, to use another phrase, in

one's "notional world",   viz.   the world we obtain by bracketing
(in Husserl's sense) all the supposed referents of mental states
in the external world,   then indeterminacy of reference follows.
Putnam also holds that this consequence entails the falsity of
metaphysical realism. Affirming the antecedent then gives the
desired conclusion. We can spell out the details of the argument
as follows (cf. Putnam, 1981, pp. 22 - 48):

(M1)   One's "notional" world does not determine reference.
(M2)   If one's notional world does not determine reference,
        then reference is indeterminate.
(M3)   If reference is indeterminate, then metaphysical realism
        is false.
(M4)   Hence, metaphysical realism is false.

By 'reference' Putnam here means the relatum of a semantic language-world relation (in something like a "presystematic" or
"ordinary" sense). I think **(M1)** has been pretty well established
by his (and others') arguments, and I will accept it here without further discussion (even if the notion of reference relation
involved in **(M1)** is far from clear).[1]
     The problems in Putnam's above argument do not lie so much
in **(M1)** as in **(M2)** and **(M3)** - and in any case I will not oppose
(M1) below. Let us consider **(M2)** first. Putnam's main argument
for its truth is the following. Given that some kind of language-world reference is needed (cf. **(M1)**), a metaphysical realist's notion of reference will have to be rather holistic and
(therefore) indeterminate. More specifically, Putnam's main
claim is that "no view which only fixes the truth-values of
whole sentences can fix reference, even if it specifies truth-
values for sentences in every possible world" (Putnam, 1981, p.
33). Now Putnam seems to think that the metaphysical realist
must accept the "received view" of interpretation which is just
a view he is referring to in the quoted sentence, viz. a view in
which one attempts to fix the intensions and extensions of
individual terms by fixing the truth values of whole sentences.
     According to Putnam's "received view" of interpretation,
the referents of terms are fixed in the context of discourse by
appeal to operational and theoretical constraints of various
sorts (Putnam, 1981, pp. 29 - 32). The operational constraints
are some kind of operational definitions. The theoretical constraints are such factors as determinism, "conservationism",

simplicity, and, I would like to add, explanatory power. The
operational constraints help to fix the contents of individual
terms whereas the theoretical ones concern the admissible formal
structures by means of which collections of terms may be related. I shall not here go into a detailed description of this
view (which, by the way, seems to me very Quinean if not, perhaps, quite acceptable to Quine).

The received view may seem plausible for a metaphysical
realist. However, I do not think that he is conceptually committed to it. For instance, a metaphysical realist could accept
conceptual abstractionism (e.g. concept empiricism; cf. Chapter
3), which is a stricter and more rigid view than the received
view. He might, alternatively, accept some kind of use theory of
meaning. Now both (some versions of) abstractionism and use
accounts may reject the assumption that intensions and extensions somehow become determined through the truth values of the
sentences in which they occur. And if they do or may do so it
remains to be shown that indeterminacy of reference follows.

How, then, does Putnam argue that if a view only fixes the
truth values of sentences then the indeterminacy of the reference of terms follows? His argument is very simple. He claims
that he can show that there are several equally adequate reference relations (satisfying all the operational and theoretical
constraints), and that none of them can be regarded as **the**
reference relation. His proof amounts to pointing out that the
so-called isomorphism theorem (well known from standard logic)
suffices to entail indeterminacy. More exactly, he treats intensions of terms as functions from possible worlds to sets of
objects in those possible worlds. Thus, the intension of 'cat'
is a function which picks out from each possible world the set
of possible objects which are cats in that possible world. These
kinds of "intensions" are of course extensions, viz. extensions
in all possible worlds.

Putnam formulates the theorem in question as follows:

"Let L be a language with predicates $F_1, \ldots, F_k$ (not
necessarily monadic). Let I be an interpretation, in the
sense of an assignment of an intension to every predicate
in L. Then if I is nontrivial in the sense that at least
one predicate has an extension which is neither empty nor
universal in at least one possible world, there exists a
second interpretation J which disagrees with I, but which

makes the same sentences true in every possible world as
I does" (Putnam, 1981, p. 217).

In Putnam's theorem the interpretation function J differs from I
only by permuting some sets of objects with the effect of creating a set of structures isomorphic to a set of structures given
by I. (Thus, e.g., the sets of cats, under I, in some structures
may become mapped into corresponding, isomorphic sets of dogs
under J, changing the interpretation of the term 'cat' nonsensi-
cally.) The above theorem is thus seen to be based on the isomorphism theorem according to which all structures isomorphic to
given models (of some sentences of the language) are themselves
models of those sentences. (Similar but still stronger claims
about reference can be based on the Skolem-Löwenheim theorem,
which allows for non-intended models of different cardinal-
ities.)
    But although his theorem is formally correct Putnam by-
passes the very central issue of showing that human languages
can be treated in this non-pragmatic way. Why should we think
that meanings of terms are extensional in the way Putnam's
"intensions" are? Why should we think the isomorphism theorem
(and the Skolem-Löwenheim theorem) applies to human languages
(including scientific ones)? I do not myself see strong reasons
for affirmative answers to these questions. The issue here
partially amounts to the much discussed problem of how to distinguish the set of "intended" models from the set of all models
of a theory or a set of sentences (see the discussion in Chapter
V of Tuomela, 1973). Furthermore, there are arguments related
e.g. to indexicals purporting to show that the formal languages
to which Putnam's isomorphism theorem applies do not adequately
model natural languages, for they lack the relevant pragmatic
elements.
    My main reason for refusing to accept Putnam's theorem
comes from my reason for accepting a use theory of meaning (of
e.g. Sellarsian type). For use theories obviously do not sharply
distinguish between pragmatic and semantic factors, contrary to
the above Putnamean semantics. Consider thus Sellarsian semantics, for instance. According to it reference is an intralin-
guistic notion, as we just saw. The Sellarsian counterpart,
pragmatic **Ersatz**-reference, of Putnam's reference is determined
by various psycho-socio-historical factors. It seems that no a
priori proof of any more the (possible) determinacy or indeter-

minacy of Ersatz-reference can be given. Whether or not people's
linguistic behavior will be found to express indeterminacy of
Ersatz-reference can only be seen a posteriori.

In any case, it seems that at least in standard cases and
under normal conditions people succeed in referring to (at least
ordinary) objects in a determinate fashion, e.g., they do not
apply the word 'cat' to dogs, and so on. Indeed, from an evolutionary point of view such determinate reference should have
survival value, and here we have a broad additional reason to
expect (a high degree of) determinacy of reference. (Cf. also
the account of picturing later in this chapter and my claim
about its determinateness.)

As to premise (**M3**) of Putnam's main argument against metaphysical realism, it requires more argumentation than Putnam
gives to it. He seems to think that, at least roughly, indeterminacy of reference (which he thinks is involved in metaphysical
realism) makes the objects of the world somehow indeterminate at
least in the sense of admitting several equally adequate descriptions of the world (cf. clause (b) in the earlier characterization of metaphysical realism). But this claim by Putnam
relating language to the world is problematic. The world might
conceivably be determinate even if our ways of referring to it
might be incomplete and indeterminate. As, e.g., Post (1983) has
argued, a metaphysical realist may then defend himself by claiming that, after all, his position does not require determinate
reference. All he needs is determinate truth of sentences, viz.
that the world (or the "intended" parts of the worlds, not
involving minds, etc.) determines uniquely the truth values of
sentences (cf. the account of picturing to be defended below).

As I do not myself defend metaphysical realism I do not
feel a need to defend the negations of the premisses (**M2**) and
(**M3**). Our above discussion should, however, suffice to show that
Putnam's central argument against metaphysical realism on the
basis of a model theoretic account of reference is full of
problems if not clearly false.[2]

Another line of argument used by Putnam against metaphysical realism relies on what Kant says for empirical realism and
against transcendental realism. Putnam (1981) agrees with Sellars' (1967b) view that Kant merely appears to endorse the
correspondence theory of truth: in fact Kant favors a strongly
epistemic interpretation of the concept of truth. Although Kant
maintains that truth is correspondence of a judgment with its

object, this correspondence is to be understood in the epistemic sense. (cf. Kant, 1787, B82 and Sellars, 1967b). In the end the objects of judgments can only be taken to be phenomena (which exist "actually" but not "in themselves"), and there is no one-one correspondence between phenomena and noumena. I take Putnam's (and Sellars') Kant interpretation to be a correct one, provided that Putnam accepts that the truth of a judgment nevertheless requires some sort of a relation between language and the (empirical) world. Yet this does not suffice to prove the invalidity of the standard correspondence theory of truth in the case of scientific theories, for they speak of scientific objects (i.e. the scientific realist's counterpart of noumena), not, strictly speaking, of either phenomena or noumena.

3. How can we show, then, that epistemic scientific realism is an improvement over its non-epistemic cousin? If we would accept Sellarsian semantics, we could present a rather short argument for the epistemic nature of realism: the concept of truth is (in accordance with the negation of the Myth of the Given) a purely language-internal one and therefore a matter of the order of conceiving; but since matters in the order of conceiving can be shown to be epistemic (the notion of a conceptual scheme is epistemic, containing or presupposing knowledge) the notions of truth as well as realism are epistemic. (See also Rosenberg, 1980, for an analogous argument.) But this argument contains rather heavy presuppositions and appears almost circular. Instead of using it we shall approach the issue from a broader perspective. I shall try to show, briefly, that human language use - and therefore also the use of scientific language - is best thought to be based on a kind of epistemic concept of truth.

Let us assume that 1) a person P claims that the earth is flat, that 2) P makes his claim by uttering the sentence "The earth is flat", and that 3) P's utterance act is a so-called literal constative speech act. All these assumptions are needed to decide whether the claim put forth by P is true or false. But if we are merely interested in the concept of truth, we shall not, it seems, need the third assumption; it is enough that "The earth is flat" is a semantically meaningful sentence of English. In any case this much must be assumed, and, in addition, that the meanings of the words 'earth', 'is' and 'flat' are known. I

shall take it that here the meanings of the words are specified
by the rules for the use of words in the language. They can be,
for instance, language-world, language-language, and world-
language rules as explained in Section II of Chapter 3. These
rules incorporate semantic as well as factual assumptions -
background assumptions and knowledge of the speaker (cf. Chapter
3). In accordance with this general view, scientific terms can
then be thought to obtain their meanings from the theories in
which they occur (cf. Sellars, 1948b, Tuomela, 1973, Chapter V).

Whether or not we want to draw a distinction between analytic and synthetic statements (i.e., between statements whose
truth is due to the meanings of the words involved and statements true due to what the world is like), we can now observe
that the meanings of linguistic items depend not just on language in the strict sense but also on their conceptual scheme
(characterized e.g. in terms of linguistic roles), on background
knowledge presupposed by this conceptual scheme, and often also
on specific scientific theories. In short, let us say that meanings are ineliminably tied to some **epistemic viewpoint**.

Now the truth-predicate applies only to semantically meaningful sentences. Thus, we can only speak of the truth of the
claim "Genes are DNA-segments" if we are aware of the fact that
the sentence is semantically significant (meaningful). Talk of
its truth or falsity does not as such presuppose that we know
the exact meaning of the sentence and are able really to **specify**
the epistemic viewpoint presupposed by it, although a justified
decision about its truth value does presuppose that. We can also
say that the predicate 'true' obtains part of its meaning from
the occasions in which metalinguistic expressions such as
"'Genes are DNA-segments' is true" are used, and these occasions
presuppose that the speaker has relevant background knowledge.

We can also approach the epistemic nature of truth from a
less linguistic and a more traditional point of view - a Kantian
point of view. Kant claimed that the idea that knowable objects
(knowable "this-suches", to adopt Sellars' (1967a) somewhat unorthodox Kant-interpretation and terminology) are located in
space and time entails that they must conform to certain general
laws. That being the case was supposed and argued by him to be
knowable a priori. Suppose thus that a person knows that something is so and so in a direct, "intuited" way, leading, e.g.,
to the representing "This is a ball". Such here-and-now knowledge obviously presupposes that, on that occasion, "This is a

ball" is a true sentence, a correct "intuitive representing".
But when the ball is thrown it still remains the same ball, we
assume. Here-and-now knowledge about the ball will soon become
there-and-then knowledge about it. But such diachronic continuity presupposes underlying knowledge, e.g., that the ball stays
the same when its spatiotemporal coordinates are changed (cf.
Sellars, 1974, pp. 336 - 338 and Rosenberg, 1980, Chapter II).

Without going into further argumentation and documentation,
let me just point out that for Kant the epistemic ladenness of
singular knowledge-claims became the claim about the existence
of synthetic a priori knowledge roughly of the form "Singular
knowledge (and truth) presupposes that the objects which it is
about obey general laws satisfying such and such conditions
(e.g. laws of continuity and causality)". Such a priori knowledge, however, is conditional when spelled out: "If there is
singular knowledge (and truth) then the objects which it is
about obey general laws satisfying such and such conditions".
While I do not think that Kant was right in regarding even such
conditional knowledge-statements as a priori in his strong sense
we need not quarrel about that here. (Note, too, that Kant
obviously thought that the antecedent can be justifiably asserted, but this at most allows one to regard the consequent as
a priori relative to the antecedent.) The basic point to be made
is that this Kantian way of approaching the epistemic ladenness
of truth leads to essentially the same conclusion as our earlier
linguistic approach. For, according to the Kantian approach,
singular truth claims (and singular knowledge claims) presuppose
general background knowledge, which includes at least synthetic,
factual knowledge (general laws of nature such as the laws of
causality, reciprocity, etc.).

We have now in fact already shown (at least relative to the
assumptions we made) that the concept of truth is in the following sense an epistemic one: **talk of truth presupposes some
epistemic point of view or other** (though not specification of
one). However, this conclusion must be qualified at once. Points
of view are mutable, and the information contained in terms of
them is transformed and renewed. How, then, can truth be in any
important sense tied to specific information? The answer is that
it is not. Although talk about truth presupposes some epistemic
point of view, it does not presuppose any very specific information. The notion of a point of view can perhaps be defined as a
suitable equivalence class of specific pieces of information

(i.e., of facts, theories, metatheories, various kinds of conceptual and metaphysical truths, etc.), and this makes truth an
idealized notion. One possibility then is to define those equivalence classes (and hence the concept of a point of view) in a
way which makes (ideal) truth "truth in the long run". In other
words, the truth of a statement is not in the end grounded on
the viewpoint adopted at any specific time but on what can be
construed from that point of view in accordance with some suitable rationality assumptions, or on what emerges from the point
of view, as it were, with the help of such rationality assumptions (see Chapter 7). The long-run point of view need not
in reality be humanly accessible, nor need truth be strictly
connected with acceptability in the long run (cf. Putnam, 1981,
p. 49, for a definition of truth as ideal rational acceptability). Thus it is compatible with what was said above that something is true (in the above epistemic sense) without its truth
being finitely or recursively knowable.

Let us note that my remarks are not, as such, directed
against the idea of truth as correspondence; nor have I here
tried to analyze away the notion of true by means of other
(epistemic) notions. Later (in Section II) I shall in fact adopt
a kind of strongly qualified correspondence view with respect to
descriptive sentences.

But what if there is, after all, an absolute Viewpoint
which corresponds to the way the world really is and which
generates the metaphysical realist's one and only True description of the world? The answer is that such a description is not
consciously accessible by human beings. We cannot intentionally
transcend our own point of view (or our points of view) in the
required way, although we can idealize them in different ways.
The notion of a unique viewpoint is not very clear (although I
maintain that it is a logically possible notion, roughly in the
way a skeptic's claim that we have always lived in a dream world
perhaps is a logically consistent claim).

On the other hand we can and do have a set of semantically
and epistemically meaningful different viewpoints and explanatory schemes based on these viewpoints. Consequently we should
look upon existential issues from the point of view of different, better or worse, viewpoints and explanatory schemes; and as
realists we accordingly claim that the scientific image gives
better explanations than the manifest image and therefore better
grounds for judging what there is and isn't.

As to Kantian lines of thought, there is no transcendental viewpoint which could justify or even make intelligible the identification of Kantian noumena with either the transcendentally ideal or real.[3] As I already said we only have the explanatory schemes accepted at a time and some possibilities to match their explanatory power with the explanatory powers of predecessor schemes. In putting forth this claim I at the same time claim that it is not possible to say, meaningfully and with good justification, anything about what the world really is like as conceptually independent of us (see below II.1. for this sense of conceptual dependence).

This does not prevent us from looking upon human beings as being under the causal control of the rest of the objectively existing world (which includes other human beings). Thus internal scientific realism can also emphasize that nature so to speak guides man, through feedback, to rectify his mistakes and, in the long run and ideally, to achieve his epistemic and other goals (especially the epistemic goals concerning the constitution of the world). However, this can never take place "outside" a conceptual scheme (although such a scheme, viz. framework for action, is under the "objective control of the world"). Nothing like "nature's own conceptual scheme", "a ready-made world", or "the language of the (Kantian) noumena" exists.

I have argued above against metaphysical realism and for the view that (acceptable) scientific realism must be epistemic and non-transcendental.[4] We can now analyze metaphysical realism in relation to the Myth of the Given. First, the fundamental assumption of metaphysical realism, that there is a ready-made world, is obviously equivalent with the thesis (MG_o) (see Chapter 3): metaphysical realism accepts the ontological version of the Myth of the Given. Metaphysical realism can however have stronger or weaker forms: the weakest one accepts only (MG_o), the stronger ones also (MG_e) (the epistemic version) or (MG_1) (the linguistic version), and the strongest forms adopt all three.

This ties up with what has been said earlier, as follows. The acceptance of (MG_o) corresponds to the transcendental assumptions of the existence of a unique ready-made World (cf. assumptions (a) and (b) of metaphysical realism). (MG_e) corresponds to the transcendental assumption of the existence of an a priori privileged epistemic (justificatory, etc.) viewpoint, call it Viewpoint. As I have thus argued that metaphysical

realism is tied to the Myth of the Given, I can draw on my
arguments against the different versions of the Myth of the
Given in Chapter 3. Especially my criticisms of theses (MG_o) and
(MG_e) apply, mutatis mutandis, as critical remarks against metaphysical realism. We can, for example, oppose the idea of ab-
stracting correspondence truth from empirical states of affairs,
an idea which would correspond to concept abstractionism (cf.
assumption (c) of metaphysical realism). (Let me here also refer
the reader to Putnam, 1981 and 1982a, where further relevant
critical remarks can be found.)

II. Causal Internal Realism

1. I can now formulate, on the basis of previous discussions,
some criteria of adequacy for scientific realism. I shall call
such acceptable scientific realism causal internal realism
(CIR). Actually, I shall define no less than a family of different forms of internal realism (one member of which is Sellarsian
realism). They are tied into a family by a set of shared conditions; as far as their other properties are concerned they can
be even logically contradictory.

We have in fact already discussed in this book the following non-transcendentally construed general - and transcendental
in a non-Kantian sense - factual "axioms" or presuppositions. I
can therefore move directly to my definition and comment on the
conditions afterwards. I shall define causal internal realism
(CIR) as the conjunction of the following principles:

(A1) There are real particulars (objects, events, processes,
      etc.) which are mind-independent.
(A2) These particulars interact (or at least can interact)
      causally with each other, and thus with human beings, in
      a way which makes, e.g., man's learning and awareness of
      the world possible.
(A3) The world, our knowledge and language are nevertheless
      not "given". In other words:
      (a) there is no ontologically given, categorially ready-
          made real world (= not-(MG_o)).
      (b) persons cannot be in non-conceptual but yet cognitive
          epistemic commerce with the world (= not-(MG_e)).
      (c) there is no conceptually privileged, i.e. semanti-

cally irreplaceable language (or conceptual scheme)
(= not-(MG_1)).

(A4) In describing the world science is the measure of what
there is and of what there isn't.

Of these **(A1)** is an ontological condition which postulates
the existence of real particulars and thus a real world which is
in a sense independent of man. **(A2)** postulates that man (as a
part of nature) and the rest of nature causally interact,
causality being here understood in a broad sense involving only
innocuous ontic import. **(A3)** denies the Myth of the Given, and
(A4) puts forth the **scientia mensura** -thesis. The conjunction of
(A1) and **(A4)** implies that there are knowable "things in themselves" (contra Kant, although not within his framework). The
assumptions **(A1)** - **(A4)** are not a priori immutable but open to
criticism and revision. In spite of their generality, their
truth is also here dependent on viewpoint and background assumptions (rather than being somehow absolute), and this in part
accounts for their openness to criticism. Let us now have a look
at these conditions, one at a time.

2. Axiom **(A1)** is fundamental for all versions of realism. The
basic problem with it is naturally its unclarity. Let us take a
closer look at what might plausibly be read into it. What is at
stake here is not how we distinguish realism from instrumentalism, for instrumentalists (at least those discussed in Chapter 4
and 5) are typically willing to grant the existence of (in some
sense) mind-independent empirical objects. Rather the crucial
issue concerns the relationship between the various versions of
realism on the one hand and realism and idealism on the other
hand.
In what sense or senses does or can realism regard some
real objects as mind-independent and in what sense all of them
as mind-dependent? These are obviously broad questions, but let
us consider some possibilities. First, as to mind-independence,
it might be proposed that the independence is of a logical
nature. As such independence relates strictly speaking to sentences only (there are no logical connections between extralinguistic entities), the thesis amounts to the claim that sentences like "Copper expands when heated" or "This swan is white"
are logically independent of sentences implying the existence of

a mind or something mental. Indeed, such logical independence obtains in the case of our above example sentences, for instance, at least under a narrow interpretation of the word 'logical', viz. logical independence in virtue of the laws of logic.

If 'logical' is interpreted broadly to amount to something like a synonym of 'conceptual' the issue is somewhat more problematic. For then we are dealing with linguistic (or other representational) entities standing for or describing some non-linguistic entities, and these descriptions of real objects should then be seen as related to some conceptual system. Now, if conceptual systems logically (in the broad sense) involve (possible or actual) thinkers and communicators, a kind of mind-dependence of these descriptions may be argued to follow, and as a consequence a kind of indirect mind-dependence of some so described real objects would be involved (cf. Kant and above Section I). Here we have a sense in which real objects can be said to be mind-dependent, then.

I shall not, however, discuss this point here further as this issue is not directly related to ontological mind-independence, which (A1) is concerned with, and, what is more important, because an idealist can accept all we say about independence and dependence within this linguistic setting. For instance, a suitable kind of idealism (phenomenalism) can accept that "There is a chair" is not only true but also in a strict sense logically independent of (sentences about) phenomenal objects (see Cornman, 1975). Thus idealism and realism are on the same footing in the case of logical independence, viz. declaring independence in the case of the strict interpretation of 'logical' and indirect dependence in the case the broad interpretation of 'logical'.

Let me, however, still take up one central point related to conceptual independence. It is that we should understand the ontological independence (A1) speaks about in relation to (A3), especially to the rejection of an "ostensive" or "necessary" connection between real objects and some relevant conceptual or linguistic entities. Thus, as argued in Chapter 3, there are no such "magical", irreplaceable ties between a language and the world - for instance, a spade need not necessarily be named by the word 'spade' (or even any of its dictionary-synonyms) in English. (If you find this talk about necessary or logical ties unclear, use the metalinguistic explication of Chapter 3 to clarify it.) We can then say that in this important sense real

objects are logically or conceptually independent of their representations. As such representations may be linguistic or mental, viz., thoughts (recall the analogy theory of thinking), we
may speak here both of conceptual independence and quite literally of the mind-independence of objects (although of course
not of the concepts of such objects). We are then relying on the
denial of the linguistic Myth of the Given. But again, all this
may in principle be available also to an idealist of a suitable
kind. We should thus look for a sense of mind-independence which
is not available to any idealist.

The basic sense in which real objects can be and are ontologically mind-independent is surely some kind of causal independence. This is at least what a realist maintains. There is
only the difficulty of sorting out the specific senses in which
such causal independence holds. There seem to be at least two
causal senses in which real objects are mind-independent and
which do not represent viable options for an idealist. I will
call these senses **causal stability** and **causal inexhaustibility**
(in terms of mental predicates). Let me briefly motivate and
explicate them.

Real objects (including human beings) and the world they
help to constitute undoubtedly causally subsist even if a) we
close our eyes and indeed even if we deprive ourselves of all
sensory information. Furthermore, at least some real objects
certainly may exist in the world even if b) all mankind is de-
stroyed and even if there had never been any human beings.
Points a) and b) must surely be accepted by anyone (except the
extreme skeptic, who has, however, been refuted convincingly
enough). Real objects thus are agreed to possess stable causal
power (cf. axiom **(A2)**), which they do not lose in the cases a)
and b). This is compatible with there being no a priori given,
categorially ready-made world (cf. **(A3)**). For, first, the world
as a whole may have such stable subsistent causal power to a
great extent independently of how exactly it is carved up or
conceptualized into bits and pieces (cf. **(A4)**). (The metaphor of
the world as a kind of metaphysical dough carvable into bits and
pieces in various ways is the metaphysical realist's and it
should be treated with proper care.) Secondly, there being such
subsistent causal power does not presuppose that the states of
affairs and events entering the involved causal relationships
can be antecedently clearly identified - rather identification
is supervenient on those causal relationships (cf. **(A4)** as

connected to the discovery of causal connections serving as a
basis of individuation and identification).

The above stability-sense of mind-independence does not
seem to be available to an idealist even if he accepts the
existence of real objects. He might possibly be able to construe
causation - as it were - within his system and claim that real
objects are not (causal) products of human minds but causally
subsistent in his newly constructed sense. Perhaps an objective
idealist (as opposed to a subjective idealist) could succeed in
something like that. But then either (i) his notion of causation
would not be acceptable to a realist or, if it were, (ii) his
position would collapse (or at least seem to collapse) into our
realism. Case (ii) follows if we require axioms (A2) - (A4)
(especially (A2)) to hold and if we analyze real existence in
terms of causal interaction. For our (A1) and (A2) together
entail that there cannot be mind-independent ideas (in an intended idealistic sense smacking of conceptual contradiction),
for such ideas would have to be real in that they would participate in causal interaction with other objects, whence the collapse. If there indeed could be such idea-objects we would here
be dealing with a version of idealism which is compatible with
realism (albeit perhaps not yet with the kind scientific realism
satisfying (A4)).

The second central causal sense in which the real objects
postulated by (A1) are mind-independent is that these objects
are not exhausted by mental properties or features. No matter
what and how many mental descriptions and attributes we ascribe
to real objects they still may transcend the realm of mental
attribution, so to speak. For instance, we may consider real
objects or phenomena inaccessible to the senses and in that
sense at best indirectly knowable (cf. Hellman, 1983). Perhaps
black holes or states of the cosmos prior to the existence of
any sentient beings could be mentioned as examples which are
impossible or at least difficult to deal with in terms of mind-
dependent predicates and descriptions.

I am not sure how to explicate this intuition, but let me
propose an ontological (and non-epistemic) principle, which
amounts to the negation of the so-called principle of the supervenience of the physical on the mental. Let us thus assume that
there we have succeeded in dividing all predicates into mental
and physical. (For instance, the construction given by Hellman
and Thompson, 1975, will do for our present purposes.) Letting ψ

be the set of mental predicates and $\emptyset$ the set of physical (or non-mental) ones our principle of non-supervenience or non-exhaustion can be stated as follows, if we allow the simplifying restriction to monadic predicates only:

(NS) $-(\forall x)(\forall y)((\forall Q \in \psi)(Q(x) \equiv Q(y))\; \triangleright\!\!\rightarrow\; (\forall P \in \emptyset)(P(x) \equiv P(y)))$

Put verbally, the principle of supervenience says that for all objects x and y, were x and y to (simultaneously) have Q or fail to have Q, for all mental predicates Q, then they would (simultaneously) have or fail to have P, for all physical predicates P. The implication $\triangleright\!\!\rightarrow$ is a counterfactual implication, representing causal and nomic connection, whenever applicable (see Tuomela, 1976, 1977). The negation of supervenience, viz. (NS), then says that even if every x and y would be mentally indistinguishable (in terms of ψ) there might still be a non-mental predicate $P \in \emptyset$ which one of the objects has and the other fails to have. So this is what the mental inexhaustibility of the realm of the physical amounts to.

Perhaps the best arguments for (NS) are only indirect ones concerned with what we can know about the physical (non-mental) features of objects suitably transcending our senses (cf. our examples of the features of black holes, etc.). Indeed, we can say in more general terms that our arguments in Chapter 4 for the explanatory incompleteness of the manifest image apply here. (Recall the aqua regia -example and the various other examples and the theoretical arguments based on them.) Accepting something like the Kantian Copernican revolution as expressed by our scientia mensura -thesis, viz. axiom (A4) of (CIR), we have a direct connection between best explanation and ontology. Thus the failure to achieve best scientific explanations in terms of ψ-predicates (and thus the manifest image) gives support to the ontological principle (NS).[5]

But even if all the above were granted would it still not be possible that black holes and other real entities transcending mental describability would be only some kind of purely spiritual or mental entities? In other words, could we not abandon our earlier intended interpretation of the ψ- and $\emptyset$-predicates and take ψ-talk after all to concern only "ideas"? Perhaps this is logically possible - but in any case I would like to retort much as above in connection with the case of causal stability. That is, such entities would then be causally

active entities presumably compatible with the scientia mensura
-thesis - and there is no need for any more a realist than a
naturalist and perhaps even a materialist to object to such
entities!

As said, condition (A1) can be considered to be the corner-
stone of all varieties of realism. It entails the negation of
strong idealism according to which only man's consciousness
exists. (A1) and (A2) together imply that there can be no proper
"idea objects" ("mental" particulars of some kind) which are
independent of human consciousness, because they would not be
real in the sense of being capable of causal interaction (which
incapability seems a reasonable assumption). In summary, while
we have above argued that the world is in an indirect sense
dependent on our conceptual sense we have on the other hand
found (at least) two causal senses serving to explicate (A1) and
making the world mind-independent.

3. Our above explication and defence of (A1) still does not
suffice to clear up all the problems it faces. Further difficulties become apparent when we compare it with postulate (A3)
(a). One could try to claim that if there are real particulars
and if the world is a conglomerate of these particulars, there
is bound to be a ready-made world in the sense of condition
(A3)(a). But this is not the case, and the problem is solved as
follows.

(A1) must be given a reading according to which the notion
of a particular is vague. I shall assume, accordingly, that the
best-explaining theory resolves not just what specific properties (etc.) real objects and the world have but also the
ontological types of these objects (cf. (A4)). In other words,
best-explaining science will decide "how the world is to be
sliced", whether the postulated real entities are objects,
events, processes, fields, or whatever, as well as what general
and specific features and properties they have. And this description or conceptualization is dependent on human consciousness in the (perhaps slightly trivial) sense that it is tied in
an idealized and indirect way (and perhaps through analogies
only) to human conceptual schemes (cf. the "internal" nature of
our realism and subsection 2 above). The notion of a particular
in axiom (A1) must therefore be understood to be indeterminate -
the exact nature of these particulars (and even the principles

of individuation) will be decided by science. (We could also
speak, in **(A1)**, of some kind of amorphic "singular mass" which
has causal powers, if this transcendental mode of speech did not
raise more problems than it solves.)

When we specify the objects of **(A1)** we could perhaps go so
far as to speak of double existential quantification. First,
(A1) quantifies over the possible ontological ways of slicing up
the world, say of "ontological articulations". Secondly we can
say that for each such articulation there are predicates which
realize **(A4)**, i.e. those in the best-explaining theory. The
problem of course is the range of the quantifiers. If the range
is defined to comprise all the articulations possible to reason
- in the sense of articulations which satisfy **(A2)** - and the
predicates of rationally possible linguistic systems, both taken
in an idealized sense, we get a realism which is relativized to
reason and rational beings (perhaps in an idealizing way) and
which also is objective. If, on the other hand, we allow wider
classes of predicates and articulations we end up with something
inescapably unintelligible (for what could such classes be?).
The kind of transcendental (metaphysical) realism resulting from
this latter possibility would be in contradiction with the
scientia mensura -thesis **(A4)** which, in the end, tells what
these classes are. (I am assuming here that the ontology specified by **(A4)** satisfies condition **(A2)**.)

In all, we perhaps ought to pack the interpretation of **(A1)**
directly into the concept of science and note that the said
articulation classes and predicate classes are exactly as much
and as little dependent on man or, better, conceptual schemes
(and reason) as is science and the existence of science. We can
also say that, given our account, there are in the world objective relationships and structures which are independent of human
mind (and reason) and of conceptual schemes - e.g. James can be
(quite objectively) taller than John, and if that is a fact, it
is independent of any conceptual scheme (although its categoriz-
ation is not).

According to **(A2)** human beings - which are in the world in
the sense of **(A1)** - can interact causally with the other entities postulated by **(A1)**. If these real entities are e.g.
objects in the standard sense, causal interaction means, more
specifically, that some events associated with these objects
affect man's physical and mental states. The problem with **(A2)**
is that it ought to be given a testable specification. The issue

is a complex one for it is connected with the problem of how man
- both as a species and as an individual - develops and learns
to cope with his environment. Although spelling out this story
is not the primary responsibility of a philosopher, something
can be said of its general features.

     Learning, in the sense of (A2), must include cognitive
learning, learning to represent the world. Thus it includes,
among other things, how persons manage to acquire true justified
beliefs or knowledge of their environment as well as conceptual
schemes which they can use in their descriptions of the environment. A theory of learning which handles these aspects adequately no doubt talks of trial and error, perhaps of insights and
the like, but in any case of the environment's corrective feedback (e.g. in the form of punishments and rewards). The entire
learning process is very complicated and largely unknown for the
time being. The same can be said at least of the cognitive
development and learning of the species homo sapiens. The idea
is, anyhow, to come to see the world as a naturalistic system
which grows representers and which accordingly comes to picture
itself within itself (cf. Rosenberg, 1974, and below Section
III).

     Let us recall in this connection Putnam's view of realism
as a theory which explains the success of science (see pp. 95 -
96). The idea can be presented in a more general form as follows
(cf. Putnam, 1978, pp. 100 - 107). The fact that human beings
generally achieve their goals can be explained quite generally
(albeit schematically and not too informatively) by referring to
the facts that (1) human beings generally achieve their goals if
they act on true beliefs, and (2) that many of such beliefs are
true.

     When this idea is applied to Putnam's realism, the expla-
nandum consists of the fact that the sciences obtain their
goals, e.g. correct explanations of laws and facts, predictions
and control over phenomena. Assumption (2) now has to do with
the scientists' belief that the terms of mature sciences typically refer, and that the laws of mature scientific theories are
typically approximately true. (Admittedly it is far from clear
how this belief is to be applied, without ending up in a circle,
to the explanation of the most general scientific theories, such
as the relativity theory, because the content of assumption (2)
is problematic in this connection.)

     Axiom (A2) can be tied in an important way to a causal

theory of knowing and hence to epistemology. It is not possible
to develop a naturalistic epistemology here. I shall nevertheless examine one central field of causal epistemology in Sellarsian terms, viz. a kind of causal correspondence theory of
truth.

Before that let us note that conditions (**A3**) and (**A4**) of
(**CIR**) do not in this connection require further comments. How-
ever, I shall later return to matters related to axiom (**A4**) (see
Chapter 7).

III. Picturing

1. The rejection of the Myth of the Given (especially of (**MG$_1$**))
entails that a person is not in an absolute and immutable logical or conceptual way cognitively connected to his surroundings
(and to the rest of the world). His ties to the world are
exclusively causal (cf. (**A2**)). Insofar as he obtains knowledge
of the real world this knowledge is based on merely causal
connections. I shall not here attempt to speculate what the
appropriate knowledge-producing causal chains are or can be.
Instead, I shall have a look at how the mentioned correspondence
theory of truth can be developed on this causal basis.

When we want to explain wherein the truth of the true
statements about the real world is based and when we at the same
time reject the Myth of the Given, it comes as natural to think
that, when properly asserted, they reflect or picture the world.
I shall assume that the following is a quite general factual
psychological hypothesis:

(**P**)   Man can in his consciousness and his use of language
         (more or less) accurately reflect the world, i.e. form
         (more or less) truthful linguistic (and mental) pictures
         (representations) of the world.

We shall call (**P**) the picturing hypothesis. As we in a
sense indicated, reflection (picturing) has to be understood, on
the basis of the rejection of the Myth of the Given, as a causal
and non-semantical notion. It corresponds to the representational concepts postulated by psychologists (e.g. in theories of
perception) but is much more general in nature. The purpose of
(**P**) is to help to explain, in part, the success of language use

and knowledge acquisition. In favorable circumstances man indeed pictures the world as (P) says, we may assume. The arguments for this are both philosophical, related to a causal account of knowledge (see, e.g., Sellars, 1979) and empirical, related to recent theoretization and research in modern cognitive psychology and especially neuropsychology (Fodor, 1975, 1984, Ullman, 1980). However, we shall not here go deeper into that, nor shall we try to spell out under what exact conditions ("optimality-conditions") the phrase "can" in (P) can be turned into "will".

How could we make (P) more precise? Let us examine the issue from the point of view of the formation of true statements. We shall assume that the "pictures" of (P) are (naturalized) sentences of some suitable human language. (Together with the analogy theory of thinking this assumption leads us to thoughts and consciousness, if the need arises; cf. (P).) As we know, Wittgenstein presented in the **Tractatus** a theory of sentences as pictures of reality. This theory presupposes an ontology of facts (and takes the notion of a non-linguistic fact to be intelligible) as well as logical atomism, but I shall rather approach the matter from another angle and explicate picture theory in Sellarsian fashion. And instead of adopting a fact-ontology I shall use the key term 'picturing' to refer to natural-linguistic objects which picture non-linguistic ones.

2.   It is not possible to give a detailed exposition of Sellars' theory here, but I shall attempt to reconstruct its philosophically relevant basic features.[6] Let us have a look at the descriptive sentence (sentence-inscription) 'James is taller than John', or rather, the written naturalistic (or natural-linguistic) occurrence of that sentence in the order of being. Let us chop it down to its natural-linguistic components, i.e., to the tokens of 'James', 'is taller than', and 'John', and examine them in this order.

Between the name-tokens 'James' and 'John' there is a naturalistic (or natural-linguistic) relation: one of them stands to the left and the other one to the right of the token expression 'is taller than'. Let us give this relation a name, 'R_1'. We shall now assume that the name-tokens bear a specific (though complex) non-conceptual and naturalistic Ersatz-reference relation to the non-linguistic real particulars James and John. This notion of reference then is not a semantic one in the

traditional sense. As we saw in Chapter 3, it is a causal
notion grounded in complicated psycho-socio-historical world-
language, language-language and language-world relations, which
in turn are grounded on the corresponding semantic rules, especially ought-to-be rules. But as this notion of Ersatz-reference is pragmatic and concerned with agents' linguistic behavior, the conditions of Putnam's isomorphism theorem, discussed
above in Section I, do not seem to apply. Thus Ersatz-reference
cannot, at least a priori, be shown to be indeterminate in
Putnam's sense.

   We shall assume, furthermore, that James in fact is taller
than John: James bears the real relation R_r to John, that is,
R_r(James, John). (Representing this state of affairs of course
requires conceptualization, e.g., of the type we have just
given. Nevertheless we are dealing with a real non-linguistic
state in the world (something which does not presuppose an
ontology of facts). We can now say, without really attempting to
give a fully explicit definition of picturing, the following:

(K)   The sentence-token "James is taller than John" **pictures
       correctly** that James is taller than John, in the sense
       (and on the grounds) that 'James' refers to James, 'John'
       to John, and that R_1('James','John') and R_r(James, John).

   We have now in fact constructed, first, one natural-linguistic relational system and, secondly, one non-linguistic
relational system. They are isomorphic with respect to their
counterpart relations R_r and R_1 (this is implied by (K)).
   Put more generally, we can say in Sellarsian terms that in
a representational system (e.g. language) a symbol (e.g. 'o')
for an object o pictures that object as P_r by virtue of having
a counterpart character, say P_1. Symbols (e.g. 'o_1' and 'o_2')
for two objects o_1 and o_2 picture these objects as related in a
certain manner, say R_r, by virtue of standing in a counterpart
relation, say R_1. We can also formulate this for sentences (as
in (K)) and say that the sentence-token 'P_1(o)' in this situation pictures o as P_r and '$R_1(o_1,o_2)$' pictures o_1 and o_2 as
standing in the relation R_r. Thus sentences (sentence-tokens)
picture in virtue of the picturing accomplished by the names
(singular terms) which occur in them.
   I shall not here give a more thorough explanation of this
picture theory, for such an explanation is not needed for the

present purposes (see also Chapter 9). In any case this theory deals basically with singular descriptive sentences containing terms which picture objects in the real world. The truth values of the corresponding descriptive existential and universal generalizations are determined on the basis of the truth values of these singular statements as in quantification theory - but only to the extent these generalizations can be regarded as truth functions of singular statements. Note that there is reason to think that (successful, correct) picturing is completely determinate at least in the case of basic statements and their truthfunctional compounds, and at least as long as no comparison of different conceptual frameworks is required (cf. subsection I.2 above). Thus no indeterminacy of reference and truth is involved in successful picturing here (but cf. Chapter 9 for the case involving comparison of conceptual frameworks).

We can now connect this theory of picturing with the theory of truth. In the case of our example sentence this is easy, since the fundamental idea is that the sentence "James is taller than John" is true **because** (in part causally because) James is taller than John, and since picturing explicates this because of -relation. We can put the matter as follows in terms of picturing-truth (truth$_p$), viz. correct picturing:

(PT)   "James is taller than John" (interpreted as a sentence in English) is **true$_p$** if and only if **(K)**.

3. Next, we shall examine the relation of picturing-truth to the more general notion of truth as correct semantic assertability in Sellars' theory. If we accept not-(**MG$_1$**), it is natural (albeit not necessary) to think that actual meaning- and truth-claims are language-internal in Sellars' sense (see the end of Chapter 3). Sellars defines such an internal notion of truth (truth$_a$) with the help of the rules of language as follows (cf. Sellars, 1968, p. 101):

(AT)   A sentence S is **true$_a$** in language L if and only if S is correctly assertable according to the semantic rules of L (and the contextual extra information possibly required by these rules).

Truth$_a$ here means correct semantic assertability. This assertability truth is an epistemic notion. One of its virtues is its comprehensiveness, for it covers not just descriptive factual statements (the scope of (PT)) but also e.g. ethical and mathematical statements (cf. "Killing is bad", "2 + 2 = 4"). Sellars' concept of semantic assertability differs clearly from the pragmatists' notion of warranted assertability as well as from, e.g., Putnam's (1981) notion of ideal rational acceptability. It also steers around many difficulties involved in such concepts (cf. Burian, 1979). This is the case basically because 'true' does not mean the same as 'known to be true' (or anything related), nor does the truth of something entail knowledge of that truth. This applies to 'true$_a$' as well. Thus 'correctly semantically assertable' does not mean the same as 'warrantedly assertable' (or anything related, such as 'warrantedly assertable in the long run'). Correct semantic assertability is not entailed by warranted assertability, for one may occasionally have good warrants for asserting but yet be mistaken. The converse does not hold either, for a sentence may be correctly semantically assertable without one's even being in a position (or even being able, in a causal sense, to be in a position) to have satisfactory epistemic warrants for asserting it.

Let us note that (AT) concerns sentence type S and not (directly) its tokens. (AT) can then of course be applied to all tokens of S. Thus if S is a token of "James is taller than John" and L is English, the claims (PT) and (AT) are more easily comparable. The following equivalence now holds, at least when S is a suitable atomic descriptive sentence-token which has come about as a result of an intentional speech act:

(E)   S is true$_p$ if and only if S is true$_a$.

According to Sellars' (1948a) terminology we in fact need in (E), as a value for the variable S, a sentence-token which is a **verifying token** and is a correct "intuitive representing" in a kind of Kantian sense, viz., a correct "here-now, this-such" representing such as "This is a red ball"; but I shall not discuss this point in more detail here.

The argumentation for the equivalence (E) is essentially based on the notion of rule-following which is not explicitly mentioned in either (PT) or (AT). The idea is, namely, that - even if our (K) is not quite explicit about this - the correctness of picturing is grounded on non-intentional pattern-gov-

erned behavior which satisfies relevant ought-to-be rules; see
Chapter 3. According to this idea the pictured objects causally
produce (or partly produce) the picturing, at least at the time
the picturing behavior is being learned. Picturing is correct if
the picturing behavior conforms to (and in a relevant causal
sense obeys) the corresponding ought-to-be rule, otherwise it is
incorrect.[7]

On the other hand we can examine the correctness of an
intentional speech act. In the situation described by (E) we
assumed that the mentioned piece of picturing behavior can here
be described as an intentional speech act. (It is important to
keep in mind that this is not possible in the case of all tokens
of pattern-governed linguistic behavior, for there may even be
pattern-governed behavior without any intentional exemplifica-
tions at all.) The correctness of an act is in a primary sense
defined as the following of the relevant ought-to-do rules (cf.
(AT)). But we can, on the other hand, anchor the correctness of
action in pattern-governed behavior, for we can redescribe the
action as an item of pattern-governed behavior; and the correctness of such pattern-governed behavior was already discussed
and defined above (cf. Sellars, 1968, p. 136). The correctness
of picturing then becomes grounded on the correctness of the
corresponding piece of pattern-governed behavior, and this correctness, in its turn, is analyzed in terms of acting on or
because of ought-to-be rules. The picturing behavior in question is, on the other hand, describable purely naturalistically
without direct recourse to intentional psychological notions
(while behavior as action does involve such recourse). In this
way we obtain a naturalistic foundation for this theory of
picturing.

As Sellars (1963a), p. 216, in fact puts it, it is natural
to take the following assumption to be constitutive of a conceptual scheme: At least in favorable circumstances the "espousal
of principles is reflected in uniformities of performance" (including centrally invariances of pattern-governed behavior). We
can make this naturalistic grounding so firm that, in the case
of singular matter-of-factual sentences, it becomes criterial of
assertability-truth and that $truth_a$ then implies (requires,
presupposes) $truth_p$ (cf. Sellars, 1968, pp. 119, 136). Thus,
given optimal conditions (viz. satisfaction of the conditions of
correct semantic assertability) for the utterance of a singular
sentence such as "This is a red ball" asserting this sentence is

semantically correct only if the sentence indeed is true in the picturing sense, this entailment being partly grounded on the causal contribution of there being a red ball in front of the speaker in such a situation.

But because picturing behavior is tied to (ought-to-be) rules the converse (the entailment of $truth_a$ by $truth_p$ in the case of atomic descriptive sentences) can also be taken to hold. For, to consider a simple example, suppose the token utterance "This is a red ball" is correct as a picture, in other words, that it satisfies a relevant ought-to-be rule roughly to the effect that it ought to **be** the case that competent language users respond by tokens of a statement having the role of "This is a red ball" to the question "What is this?" when confronted with a red ball. But this is also what it is correct to **do** in this situation: any competent language user ought to describe the object in front of him just so. It is thus a semantically correct act to utter the mentioned sentence in this situation. Hence "this-such" responses which are correct as picturings are also correctly semantically assertable and true in that sense. Thus, in all, we end up with (**E**). (The implication from $truth_p$ to $truth_a$ is rather problematic, however, and would deserve a more detailed investigation.)

(**E**) gives a naturalistic foundation for truth. It is worth stressing that picturing is a non-conceptual notion which, however, is centrally related to a conceptual framework, as (**E**) indicates. $Truth_a$ and $truth_p$ are clearly non-identical, for $truth_p$ is a causal non-conceptual relation between the world and behavior, while $truth_a$ is not a relation at all but rather a scheme or ground for classifying linguistic expressions (sentences). Picturing (but not assertability) can be more or less adequate and accurate (in, roughly, the sense in which maps can be in varying degrees adequate and accurate).

The correctness of picturing truth can be examined from another angle too. For pattern-governed behavior can be assessed with respect to its evolutionary "value". We can have both biological and cultural evolution. From a biological point of view we are cognitive map-makers (or picturers) as the map-making ability has had survival value. Perhaps it is in part genetically determined (comparable to the case of bees and their "dances" or, closer to home, sensory - motor schemes understood in something like Piaget's sense). But we may also speak of cultural evolution. Thus we can examine e.g. how well some piece

of pattern-governed behavior (e.g. a deep-rooted manner or
habit; cf. Kuhn's notion of paradigmatic "exemplar") serves the
purposes of, say, constructing true and explanatory scientific
theories.

4.   We shall briefly return to the above action-oriented theory
of truth in Chapter 9. Let me merely note here that I have
presented an epistemic notion of truth which nevertheless does
not presuppose the linguistic version of the Myth of the Given
(MG_1) and which can be connected to a (causal) correspondence
theory of truth (in the sense of the equivalence (E)). Let us
also observe that although truth$_a$ is a language-internal affair
it can be shown to be formally compatible with the notion of
truth defined by Tarski (cf. Sellars, 1968). This holds at least
if we accept a kind of substitutional quantification with respect to an ideal language where all objects have names. But, as
far as I can see, we are able to use, via the equivalence (E)
and by assuming idealizingly that all objects have names, objec-
tual quantification, too, and hence language - world relationships - although we have here psycho-socio-historical connections rather than traditional semantic ones.
    Although the above Sellarsian account is formally compat-
ible with Tarski's definition of truth, it is stronger. As we
know, Tarski's account is philosophically noncommittal while the
above account relies on a causal view of knowledge (which, however, rejects (MG_e)). Tarski's basic notion is that of satisfaction, viz., that of a sequence of objects satisfying a formula.
But those objects could be purely conceptual objects - they do
not have to be real ones. Furthermore, satisfaction is a purely
formal notion which is not required to express specific causal
(or other factual) relations between real, extralinguistic objects and natural-linguistic objects, as does picturing. Thus
the above picturing account is able to characterize the notion
of factual truth which Tarski's account (as such) fails to do.
For his theory simply fails to explicate truth as correspondence between language and the world - it does not even aim at
it, as it takes the atomic or elementary satisfaction relation
as a primitive notion (cf. Putnam's, 1983, 1984, acute criticisms) and, what is more, as a primitive reducible to something
non-semantic. Tarski's account focuses on the recursive truthfunctional definition of truth on the basis of primitive satis-

faction - but a picturing theorist need of course not accept
such a recursive characterization for all relevant compound or
general statements (cf. ethical principles), although he may
accept it for e.g. simple factual descriptive generalizations.[8]

Hypothesis (A2) of (CIR), which has to do with causal
interaction, and especially the picturing theory contained in
it, occupies a central position when we attempt to answer why
science rather than, say, magic or religion, gives ontologically
the best, and in that sense the right, picture of the world. For
I claim that the world's own contribution constitutes an essential part of the solution of this problem, and that the right
way to take it into account is via (cybernetic) self-correcting
causality - just as (A2) and (CIR) do; and I claim that science
gives the best account of causal interaction (see e.g. the arguments in Tuomela, 1976). I also want to argue, on the basis of
what was just said, that if causality is allotted such a central
role, we need something like a picturing account of knowledge
and truth. More generally, all this is included in the scientific method, and we can thus claim that it is precisely the
self-corrective method of science which makes science the best
means of attaining knowledge (for example, art or religion are
not self-corrective in this sense; cf. Chapter 10).

CHAPTER 7

SCIENCE AS THE MEASURE OF WHAT THERE IS

I. On the Various Kinds of Scientific Realism

1. We have frequently claimed in this book that (internal)
scientific realism incorporates the idea that science is the
measure of everything that exists. Part of the philosophical
grounding for this thesis lies in the view that in principle
reason can grasp the world - and that the scientific method is
the best explicate of reason.[1] Given this, both agnosticism (the
view that the world cannot be known) and mysticism (the view
that there is some mystical non-rational method, e.g. divine
revelation, which only can help us tell what the world really is
like) become blocked.

Our general view receives support from the epistemic character of truth (cf. Chapter 6), viz. from the fact that truth is
necessarily relative to a point of view and to background assumptions and knowledge. Accordingly, as our criterion of truth
we will then have the epistemically best (or best-explaining)
theories - and of course such a true (true$_a$, that is) theory
will tell us what there is in the world. The scientia mensura
-thesis should be interpreted so as to allow for the possibility
that even best-explaining theories may involve idealized concepts and idealizing assumptions. Another qualification needed
is that no scientific theory with a finite conceptual basis
(viz. with a finite number of primitive predicates) can exhaustively characterize its domain, for all real entities can be
taken to have an inexhaustible and indefinite number of features
and aspects - for they can always be approached from new points
of view. (This is entailed by the rejection of the Myth of the
Given.)

The scientia mensura -thesis speaks about best-explaining
theories rather than about currently accepted theories. Thus it
cannot be shown incorrect on the basis of the (alleged) fact
that all or most current theories are false. The scientia mensura -thesis does not involve the claim that the best-explaining

theories will ever be found during the future history of mankind
nor the claim that they are humanly attainable even in principle. Thus this thesis should rather be regarded as a kind of
regulative standard, which allows that at any moment new conceptual systems and new ways of looking at problems may be found.
In this sense **scientia mensura** makes room for creativity and
innovative conceptual change.

Suppose now that science can indeed be regarded as the
measure of what there is. This can still be interpreted in a
variety of ways. For instance, in the terminology of Chapter 4,
it can be understood as, e.g., strong (exclusive) realism or
weak realism. Recall that weak realism does not claim that
Eddington's observational table does not exist. Weak realism is
thus compatible with the view that there are at least two adequate descriptions of a table, of which, to wit, the one in
accordance with the scientific image is the better-explaining
and generally better one. In contradistinction to this, according strong scientific realism, represented e.g. by Sellars, only
the table in the scientific image really exists. The table of
the manifest image is only phenomenal (and does not really
exist). This view of the things in the manifest image is largely
Kantian. While it is not necessary for the purposes of this book
to choose between weak and strong realism (recall that our
arguments for realism have been for weak rather than strong
realism), it is still instructive to air these issues here and
to discuss also some other variants of scientific realism.

I shall start to map the situation by classifying the various kinds of scientific realism in finer detail than I have done
so far. I shall focus on epistemic (or internal) variants of the
doctrine to the exclusion of others, relying on the stand (argued for in Chapter 6) that all defensible forms of realism are
to be found among them. The fact that scientific realism must be
scientific or be based on the method of science derives to a
large extent from the supposed nature of the scientific method -
that it is a method which can be said to deliver more reliable
and explanatory knowledge than any of its alternatives (cf.
Chapters 9 and 10 as well as Section II of Chapter 4).

In what follows I shall distinguish between some forms of
scientific realism that we have not previously discussed in this
book (cf. Cornman, 1975). As a criterion of what there is we
shall in this classification again take best scientific explanation. This is an epistemic notion and, on the face of it,

strongly idealized. The concept of explanation is a complex and many-tiered one, and it is not an easy matter to characterize best explanation. I shall return to the question in Chapter 9 and use this notion, until then, as if it were relatively un-problematic. (An adherent of metaphysical realism can read the following characterizations as a series of contingent claims about realism rather than as definitions.)

We can call minimal scientific realism (**MSR**) the doctrine which holds that science is able to describe the world correct-ly. I shall formulate it as follows (cf. Cornman, 1975, pp. 265 - 266):

(**MSR**)   All sentient and non-sentient physical objects have at least the constituents and properties which correspond to the scientific terms needed in the theories which best explain the overall behavior of those objects.

(**MSR**) thus allows both sentient and non-sentient beings to have also constituents and properties which are not postulated by best-explaining science at all. Consequently (**MSR**) reckons with the possibility that sentient beings sense colors even if the best-explaining theory does not postulate (or will not postu-late) colors and the sensing of colors. (But note that (**MSR**) in conjunction with (**A2**) of (**CIR**) - see Chapter 6 - still comes to exclude spirits, ghosts, demons, and the like spooky entities.)

Next we define a more demanding version of realism which we shall call moderate scientific realism (**MOSR**):

(**MOSR**) (a) Minimal scientific realism is true; and
          (b) non-sentient physical objects have only the constitu-ents and non-relational empirical properties which correspond to the theoretical scientific terms needed in the best-explaining theories.

(**MOSR**) thus says that non-sentient beings are "mere" scientific beings in the sense that they have precisely the constituents and properties which the best explanation of their behavior assigns to them.[2] The thesis of extreme scientific realism (**ESR**) generalizes this to encompass all real beings:

(**ESR**)   All non-sentient physical objects and all sentient ob-jects (including persons) are "mere" scientific objects.

The word "mere" is to be understood in the following way. (**MOSR**) can be abbreviated as the view that (i) all non-sentient physical beings are mere scientific objects and (ii) all sentient beings are at least scientific objects. When we strengthen (ii) we get (**ESR**) from (**MOSR**). Note that both (**MOSR**) and (**ESR**) appear, at least at first glance, to deny the existence of colors and even the sensing of colors in the physical world. When these are included we end up with what I shall call Sellarsian scientific realism (**SSR**):

(**SSR**)   All sentient beings (persons included) and all non-sentient physical objects have precisely the constituents and properties which correspond to the theoretical scientific terms needed in the theories which best explain their overall behavior, where the explanations also cover the sensory features of the world and man's reactions to them.

Thesis (**ESR**) seems to be more demanding than (**SSR**) in that it does not guarantee the describability of the sensory features of the world. It is fairly obvious that a scientific realist who emphasizes the importance of the manifest image will find Sellarsian realism more to his taste. It ought to be noted that it is possible to give (**SSR**) an interpretation which makes it equivalent to (**ESR**), but I shall not deal with this possibility here (but cf. Tuomela, 1977, Chapter 1).

While discussing the various kinds of realisms, I have all the time assumed that the theoretical terms mentioned in the best-explaining theories really manage to refer (in some suitable sense) to the said constituents and properties (cf. argument (**WR3**) of Chapter 5). In fact the realist interpretation of the theoretical terms presuppose that they can be employed in singular descriptions too - in principle at least, if not in actual practice.

All of the discussed theses refer to the best explanation of the overall behavior of the objects in view. The term behavior is to be understood very broadly, so as to include e.g. the movements of a stone, the pressure of a quantity of gas, the epistemic state of a person, etc. Cornman (1975), on the contrary, speaks in this connection only of observable behavior. But restriction to observable behavior is not right if we intend to solve ontological questions by reference to best explanation

- Cornman's theses can lead to one-sided and misleading claims about ontology.

There is no need here to go into a detailed discussion of the various versions of scientific realism presented (cf. Cornman, 1975). They represent, in any case, various interpretations of the **scientia mensura** -thesis, interpretations open for an epistemic scientific realist. To put it quite generally, we can say that they are presuppositions which underlie scientific inquiry, or factual hypotheses rather than a priori truths - thus feedback from scientific inquiry can affect their acceptability.

2. How do the above theses relate to weak and strong scientific realisms, the views discussed earlier in Chapter 4? The answer seems to be, at first glance, that **(MSR)** corresponds to weak realism and **(SSR)** and **(ESR)** to strong (or exclusive) realism. **(MOSR)** in its turn represents strong realism with respect to non-sentient beings (such as tables) but weak realism with respect to sentient beings. But this answer is incorrect. To see it, let us have a look at Eddington's table. The observable table can be, say, brown and hard-surfaced. Do the above brands of realism allow tables to have such observational properties? **(MSR)** does, but only on the additional condition that observational properties are compatible with the constitution and properties of the scientific table. Let us call this extra assumption the compatibility assumption. Then **(MSR)** together with the compatibility assumption represents a case of weak realism.

How about the other variants of realism? In order that both the observational and the scientific table can exist, it is required that the observational table is identical with the scientific table. We can say, then, that each one of the variants of scientific realism represents, in conjunction with the identity hypothesis, weak realism. In contrast to this, **(SSR)** and **(ESR)**, together with the negation of the identity hypothesis, represent strong realism, just as **(MOSR)** does with respect to non-sentient beings. (Cornman, 1975, discusses these issues in more detail; cf. also my criticism of Cornman in Tuomela, 1978a.)

It does not make a great deal of difference, relative to the view developed in this book, which one of the theses **(MSR)**, **(MOSR)**, **(ESR)**, and **(SSR)** is accepted. Perhaps we ought to say, nevertheless, that **(ESR)** and **(SSR)** appear to be the best expli-

cations of the scientia mensura -thesis in the sense that they
cover all of the properties of the objects. (In fact, (**MSR**) and
(**MOSR**) are inconsistent with the strict interpretation of the
scientia mensura -thesis according to which science tells us
exactly what there is.)

As we will see in Section II below, the scientia mensura
-thesis deals with the world only at the ontological level, i.e.
from the point of view of the causal order of being. It follows
that prescriptive matters (e.g. values and norms) remain, in an
important sense, outside it. Let us also note that the scientia
mensura -thesis accepts other good descriptions of the world,
such as those drawn by art, as illuminating. But it does nevertheless claim that when it comes to explaining matters of fact
science has the best and final say.

II. Ontology and the scope of the scientia mensura -thesis

1. All varieties of scientific realism, including causal internal realism, are based on the scientia mensura -thesis. It is
important to see precisely what the thesis entails and what it
does not entail. We have in this book leaned heavily on the
timehonored distinction between the **order of being** and the **order
of conceiving**. The distinction is similar to Kant's quaestio
facti - quaestio juris -distinction. (Cf., e.g., Kant, 1787,
B116ff, also cf. his constitutive-regulative -dichotomy, see
Kant, 1787, B222 - 223, B537 - 538, B597; we can talk also, in
this connection, about a general theoria - praxis -distinction.)
The order of being has to do with the being of real things, in
other words, with the "descriptive" content of the world. The
scientia mensura -thesis applies to the order of being and only
to it, thus making ontological claims. There are reasons to
claim that the order of conceiving is **not** so related to ontology, to what there really is. We shall now take a closer look at
these questions, starting from the order of conceiving.

The order of conceiving (here thought of as a type of discourse) deals with questions of meaning, justification and the
foundations of knowledge, as well as with prescriptive matters
(and hence with values and norms). The crucial observation is
that the order of conceiving has to do with matters which differ
from those dealt with by the order of being. For instance the
metalinguistic meaning postulates of the order of conceiving,

such as "'Bachelor' means an unmarried man", are not descriptive
(or at least merely descriptive), nor are they part of the discourse about the order of being at least in the strict sense in
which the claim "Turning on this switch makes the light come on"
is (cf. also Sellars, 1963b, p. 451).

The order of conceiving cannot be reduced naturalistically
into the order of being in the sense that the concepts of the
former cannot be analyzed in terms of those of the latter. To
assume that such a reduction could be accomplished would be to
commit a kind of naturalistic fallacy. There is thus no way in
which e.g. meaning postulates, prescriptions ("If A, one ought
to do B") or justifications ("The item of knowledge of T justifies accepting claim V as knowledge") could be analyzed (without remainder) by, say, naturalistic causal laws concerning
linguistic and other behavior (see, e.g., the arguments in
Sellars, 1963b). As we have seen, it comes as natural to think
that the conceptual order as such makes no ontological commitments, even if we admit that the order of being can bear an
"evidential" relation to the order of conceiving (for arguments
see e.g. Sellars, 1963a, Castañeda, 1975, and Mackie, 1977). But
we can yet claim that, although the mentioned prescriptive
claims and meaning postulates cannot be given an analysis in
merely naturalistic (or, more generally, descriptive) terms,
obligation and meaning are, from the point of view of the order
of being, entirely naturalistic matters. We can also say, along
with Sellars, that obligation is causally although not conceptually reducible to being.

We cannot here go into the intricacies of the detailed
relationship between the two orders. I have introduced the distinction mainly to illustrate the domain of application of our
motto, i.e. the scientia mensura -claim, for that domain is the
order of being. There are good reasons for claiming that according to the brand of scientific realism I endorse, naturalistic
(even materialistic) ontology suits well, indeed best, also the
humanities and the social sciences; we shall return to this
shortly.

Let us note, however, that although e.g. speaking (as activity) falls into the domain of scientific description, science
cannot naturalistically reduce the meaning (content) of a speech
act - precisely because semantic categories are nonreducible. On
the other hand this admission entails no commitment to any new
ontological content ("spirits" or the like) simply because sem-

antic categories carry no such ontological commitments. The same
applies to the prescriptive uses (and token utterances) of normative and axiological discourse (see Tuomela, 1984, Chapter
1).[3]

Since scientific realism in the form presented above does
not concern - to use a general characterization - the appropriate use of prescriptions, it can adopt different views as to
their nature and precise content. Thus scientific realism has,
as such, nothing to do with scientism, if by scientism we mean
the optimistic view that science can solve the central social
and technological problems. Such scientism is above all (but not
merely) an axiological view which therefore falls outside the
scientia mensura -thesis and scientific realism. It seems to me
that there are good reasons for rejecting such scientism.

It deserves to be emphasized that the distinction between
descriptive and prescriptive language use (viz. **singular** uses
and token utterances) is a fundamental one. To be sure, a great
many natural language expressions can be used both in descriptive and prescriptive contexts, and there is no bifurcation of
the vocabulary of a language into a factual and a value component. Natural language is no doubt in this respect value-laden,
but this fact by no means nullifies the above distinction and
the possibility of making such a distinction in the case of each
particular context of use of language.

As we already saw in Chapter 1, the traditional division
into the natural sciences and the Geisteswissenschaften (the humanities and social sciences) is not an ontological one. Nevertheless, it has often been (incorrectly) thought that the
Kantian distinction between the empirical and the transcendental
ego incorporates an ontological mind-body -distinction. But the
notion of transcendental ego is best thought to be a moral
rather than an ontological category in the above sense. Man has
no deep, hidden metaphysically (ontologically) important essence
which would set him apart from other real beings, such as apes,
amoebas or even stones (cf. Rorty, 1979b, Chapter VII, who
brings this point well home). Consequently, we can, for instance
accept the hermeneuticists' view of man as **pour-soi** (to use
Sartre's terms), as a being which defines and evaluates himself
afresh and which changes himself, etc., and not merely **en-soi**.
This can be done without commitment to ontological idealism; I
thus claim that hermeneutical materialism is a viable philosophical position.

To put it in slightly other words, our causal internal realism is "tough" or "strict" (i.e. materialistic) in its ontology, but "soft" with respect to concept formation, epistemology, and prescriptive matters. As far as social issues are concerned, our realism allows for **Ideologikritik** - in fact it tallies with much of the critical theory of the Frankfurt school. (However, the compatibility does not extend to methodology, because the critical theory adopts an instrumentalist view of theories.)

2. If we accept the **scientia mensura** -thesis and explicate it in the direction of, say, minimal, moderate, Sellarsian or strict scientific realism, is there anything of ontological importance yet to be added? In a sense the answer is negative, for we do not as yet know what future science brings (or can bring) with it. We can nevertheless do some guesswork on the nature of the "right" ontology. Similarly, we could try to assess the acceptability of the **scientia mensura** -thesis as a criterion of what there is. If, for instance, we could give an a priori demonstration to the effect that some ontological affairs are bound to stay beyond the reach of science, we of course would have derived a refutation of the **scientia mensura** -thesis.

Could there be something in the world which in principle escapes the method of science (and thus the best explaining theories it produces or can produce)? Suppose a critic claims that, for instance religion, magic or art (etc.) should rather be regarded as the criterion of truth. But then the reply is that none of these other methods for gathering descriptive knowledge about the world are superior to science as to the features of objectivity, criticalness, testability, self-correctiveness, autonomy, and progressiveness (to name a few central aspects) and therefore as to their power to lead to best-explaining and truthlike accounts of the world. (See Chapters 9 and 10 for an elaboration of these features.)

So let us suppose, for the time being, that the superiority of the method of science over all the other methods of gathering knowledge about the world can be granted. A critic may still claim that science does not suffice to capture all the truths about the world. Consider thus the following argument (based on an example due to Smart). Suppose we have a theory T which accords with the solid core of current science in postulating that all physical entities belong to a four-dimensional space-time manifold. Consider then a competing theory T' which is alike T in what it says about the contents of our space-time manifold but which asserts that our space-time manifold is a cross sec- tion of a five-dimensional manifold which contains as another cross-section another space-time manifold (also a product of the Big Bang!), containing, like ours, galaxies, stars, planets, etc. Let us suppose that according to T' this other sub-universe and our sub-universe are precluded by the laws of nature from causally interacting with one another so that we cannot have causally obtained knowledge about the other sub-universe (cf. (**A2**) of (**CIR**)). Now the argument is that T will be the best-explaining theory because of its simplicity. Thus, if T' after all were the true theory - as can consistently be assumed, it seems - this might seem to be a counterexample to the **scientia mensura** -thesis (understood in a stricter sense than that of (**MSR**) of Section I).

But is it really a counterexample? First, I think simplic- ity is not directly a component of explanatory power - so that does not count. Secondly, we may claim that, on the contrary, T' can be consistently assumed to give a better explanatory account of the world than T. For T does not explain anything in the other sub-universe whereas T' does. Is that not enough to rebut the critic's argument? To be sure, we do not have direct causal knowledge of how T' fares with respect to that other sub-uni- verse (nor vice versa concerning its inhabitants' knowledge, if any, relative to our sub-universe). But it may well be the case that the claim about the superior explanatory power of T' can be backed by e.g., theoretical evidence, viz. evidence based on the connections between T' and other, corroborated theories. Prob- lems related to information gathering and to the testability of theories should anyhow be distinguished from those related to their explanatory power. So it is quite reasonable to claim that T' is superior to T in explanatory power, correctly understood, and this fact of course keeps the **scientia mensura** -thesis standing. Thus, in all, the present purported counterexample has more to do with ambiguities in the notion of explanatory power and with ascertaining the explanatory power of theories than with the **scientia mensura** -thesis, and it does not qualify as a counterexample to it. (See also our discussion in Chapter 9 claiming that the maximally informative true theory will amount to the best-explaining theory.)

3. Could there be other kinds of complaints against **scientia mensura**? It might be thought that mental phenomena and features could present **scientia mensura** with difficulties. Let us now discuss these matters starting with materialism. How closely related is ontological materialism with our **scientia mensura** - thesis? In a sense materialism no doubt fits the ontology of scientific realism. But the relationship is not one of implication - although this depends somewhat on what materialism is taken to mean. (Indeed, it may even be conceptually possible to reconcile scientific realism with a kind of psycho-physical dualism - as long as the mental entities are postulated by science.)

Defining the notion of matter is no easy thing. We cannot say, for instance, that only things which have mass are material (a counterexample is provided by electromagnetic fields). Nor is the criterion provided by mind-independence, for also objective idealists postulate ideas which are, in a sense, mind-independent. (See the analyses by Cornman, 1971.) Perhaps we ought to concede the impossibility of giving a definition of materialism exact enough for philosophical purposes. Perhaps we ought to rest content with an a posteriori characterization and say: **Scientia mensura** - the ontologies of the best-explaining theories define the notion of matter. The trouble with this definition is that, arguably, it sets us to walk along a circle: we are not defining science by purely methodological criteria but in part also by substantial ones, and science seems to be connected with materialistic ontology (even dualists often admit this claim).

One aprioristic possibility remains, and that is to define matter on the basis of causal influence. We could then say that an entity is material if and only if it takes part or can take part in relations of causal interaction (or if its parts can participate in such relations); cf. axioms **(A1)**, **(A2)** and **(A4)** of **(CIR)**, and Bunge, 1981. To clarify this proposal I merely note that the qualification in parentheses is needed in order that the world as a whole (as a system) could be thought to be material. Materialism defined in this way is incompatible with both objective and subjective idealism. The former view claims that an entity is real if and only if it is (or expresses) some kind of objective and non-personal "pure" spirit (cf. Hegel's absolute spirit). According to the latter view reality has to do, entirely or at least in part, with the states and contents of a non-material human mind or consciousness.

It would, I think, be natural and justified, too, to adopt the above characterization of materialism and notice that future science (the best-explaining theories) will in principle sort out the details. There are, however, plenty of traditional arguments against materialism which, usually, are also arguments against the **scientia mensura** -thesis.

We noticed in Chapter 2 that the difficulties with the incompatibility of the manifest and the scientific images can be put into three categories, viz. the problems brought about by sense qualities, the intentionality of cognitive processes, and features of personhood. We might perhaps have added the problem of meaning and universals, although it is implicitly included in the second group. Examining the a priori difficulties with materialism we can again note that they are, too, closely tied with the mind-body -problem. (In fact I know of no serious difficulty that does not relate to the mind-body -problem). Let us note here that materialism is today the favored variant of monism and indeed the favored mind-body theory among professional philosophers. Actually it is rather difficult to find supporters of ontological mind-body -dualism among the most outstanding philosophers. Nevertheless the following traditional arguments come mainly from the dualist camp.

In medias res, thinkers from Plato and Aristotle on have assigned human minds various properties that can be regarded or are regarded as non-material. As e.g. Rorty (1979b) has demonstrated, the mind-body problem has during the course of the two millennia several times adopted a new guise; the criteria of matter have varied accordingly. To get a picture of the complex mind-body -problem I shall present a representative list of some of the central features which have been used in descriptions of the mental aspects of people (cf. Rorty, 1979b, p. 35 and Feigl, 1967, p. 29):

(i)   incorrigible self-awareness
(ii)   ability to sense qualities
(iii)   ability to exist separately from body
(iv)   non-spatiality
(v)   ability to be aware of universals
(vi)   intentionality
(vii)   ability to use language
(viii) ability to act freely
(ix)   ability to be "one of us" and to act morally.

Features (i) - (iv) form a class which in a sense generalizes
the category of sensory qualities presented in Chapter 2. At
least if they are understood through Descartes' way of thinking,
we can say that they have to do with mental substance and its
properties. In Descartes that means thoughts, experiences and
feelings, etc., which occur in the stream of consciousness.
These occurrent episodes have the property that they would
exist even if we thought that everything else (such as matter)
were annihilated or otherwise reduced away. Features (iii) and
(iv) are of course connected with the idea of such mental substance, whether it consists of changing processes ("the ghost in
the machine", as in Descartes) or of some immutable substance
(cf. the Greeks' noûs, and the soul of christianity). But also
(i), incorrigible awareness of these experiential states and
processes, is central. In its essentials it expresses the epistemic version of the Myth of the Given (cf. thesis (MG_e) and
especially (GIA) in Chapter 3). Many modern proponents of the
Cartesian line of thought (e.g. Kripke) have taken the constitutive feature of the mental, and the central problem for materialism, to be in the sensing of qualia (feature (ii); e.g. the
quality of headache), but there have been convincing replies to
this challenge by proponents of functionalist mind-body theories
(such as Rorty and Shoemaker).

It is important in this connection to notice that even if
the sensory (or other mental) qualities could not be accounted
for in a strictly materialistic way without postulating any
entities or episodes in the scientific image to correspond to
such features of the manifest image, materialism in a weaker
sense satisfying the scientia mensura -thesis would still be
defensible.[4]

I cannot here give a detailed defense of my stand, but I
would claim that no one has been able to give a clear and
coherent characterization of the notion of (purely) mental substance (or state, process); see Rorty (1979b), Ryle (1948), and
Wittgenstein (1953). As especially Ryle and Wittgenstein have
shown, notions such as a Cartesian private mental substance have
often been formed by committing a category mistake. If a person
has a headache or if he sees an afterimage we cannot conclude
that there are in these cases separate ontologically significant
entities, headaches and afterimages even within the ontology of
the manifest image (that we are presently concerned with). We

would otherwise erroneously postulate a mental state or activity to be an entity - we would perform an unwarranted transition from adjectives and verbs to nouns. It is obvious that this line of thought leads into a conceptual morass.

Such entification is erroneous and the error arises in parallel to a view of how abstract singular conceptual entities are formed. Thus here the concept of headache (of headache as an entity, a particular) has been constructed artificially from the (universal) concept of aching state, just as the concept of redness sometimes is taken to be formed from the concept red. In this way "pure spirit" (or whatever label is given to the result of the entifying operation) is as if formed of the same stuff as universals (cf. objective idealism). But if we pursue the task of philosophical analysis further and think that in the end universals are linguistic entities (or at least analogous to linguistic entities), as we did in Chapter 3, we will end up equally well with the conclusion that there are no purely mental entities (one having e.g. the properties (iii) and (iv)). For we have seen that semantic categories carry with them no ontological commitments (also cf. what follows).

Properties (v) - (vii) have traditionally been associated with questions of human intelligence. Thus the problem of the ability to have direct awareness of universals - and ability to grasp universals - was pivotal for both Plato and Aristotle (cf. again (GIA)). Note specifically that Aristotle's nôus, intelligence or reason, was in a sense a faculty separate from body (cf. (iii)), but nevertheless in many respects totally non-Cartesian; see e.g. the perceptive remarks by Rorty (1979b). We can, incidentally, take it that the medieval and modern Christian notion of the eternal soul - the notion of the "invisible man" which survives the burial of the body - grew in the middle ages from the notion of reason in antiquity.

It is of some interest to note that Greek philosophy, Aristotle's own philosophy in particular - did not recognize the mind-body problem in the Cartesian form, that is, in the form of purely mental entities which have special phenomenal qualities and which are in some sense self-verifying. For Aristotle the counterparts of these Cartesian entities (e.g. headaches) belonged squarely to the body and were therefore material (see Matson, 1966). Perhaps the most central features of the mind-body -problem, i.e. (i) - (iv), are historically rather recent and derive, for the most part, from Descartes.

To return briefly to the category of reason, we have already examined the problem of intentionality (in Chapter 2) and philosophical issues about language (in Chapter 3). I want to emphasize here that there are very good reasons to think that intentional concepts are functional - in fact Aristotle may be considered a functionalist in the modern ("Putnamean") sense. These concepts can be explicated, following Sellars, by using the concept of linguistic role (and ultimately by reference to language use), and the same recipe is appropriate for the problem of universals. Nothing mental or non-material is associated with these notions. They have to do with language use and at most with thought episodes construed along the lines of the analogy theory of thinking. Ontologically these episodes can be regarded as neurophysiological or naturalistic, anyhow. (See also Rorty, 1979b, for similar remarks.)

Properties (viii) and (ix) refer to the category of personhood in the sense discussed in Chapter 2. If we accept the solution given there, these features are to be analyzed by reference to prescriptive discourse, and no ontological worries are heaped upon materialism (provided, of course, that prescriptions involve no ontological commitments).

As these admittedly scant remarks indicate, a materialist who accepts the distinction between quaestio facti - quaestio juris (or order of being - conceptual order) in the above sense, and who, even in a more important sense, analyzes universals (mathematical universals, such as sets, included) in linguistic terms, seems to steer around the troubles involved in the mind-body -problem. Universals have given rise to a plethora of conceptual muddles in the history of philosophy (cf. Hegelianism), and one ought to take all necessary care in handling them.[5]

As to a constructive view on the mind-body -problem, I am inclined to favor a sort of emergent materialist position - which, if not correct, is at last a promising candidate for further developments (see Bunge, 1980, for this option). Its leading idea is that man is a many-dimensional material system which interacts in many ways with its environment and which is in several ways emergent relative to neurophysiological microentities as presently conceived. This matter-energy system can be described by material (or more generally, non-psychological) as well as by psychological predicates (and in some sense needs both kinds of predicates for its satisfactory description).[6] My conjecture is that future neuropsychology will merge with psychology in such a manner that it nevertheless contains counterpart predicates to today's psychological predicates.

In any case we can conclude that, as long as we reject the linguistic Myth of the Given ($\mathbf{MG_1}$), the application of psychological predicates has no impact on our materialist ontology. In other words man, as a matter-energy system, is in part describable by using (present or tomorrow's) psychological predicates - and, so it seems, is not adequately describable as a person without them (or their counterparts within the scientific image). It is important to see that - because we have not, in our Sellarsian semantics, directly connected language with ontology - we have here a view which does not, and need not, reduce away mentalistic (intentional and other) descriptions and which yet is compatible with ontological materialism.

Whether folk psychology will turn out to be approximately true or clearly false will not affect our view. If it should turn out to be true and if it could be regarded as a best-explaining theory - which both assumptions seem very questionable - there would be, e.g., states of believing in an ontic sense - but they could still be considered material (in our sense) rather than some kind of spooky, purely spiritual states. If folk psychology - and best-explaining "mentalistic" psychology - is clearly false, then, given scientia mensura, there are no specific mental states in its sense.[7]

Given the discussions above and in the preceding chapters we can conclude that we have managed to find support for the following metaphilosophical theses:

(1)   Empiricism and empiricist instrumentalism as ontological, semantic and epistemological doctrines are incorrect views (Chapter 6).

(2)   Idealism is an erroneous ontological doctrine (cf. (CIR) and Sections II and III of Chapter 6).

(3)   Metaphysical realism is an incorrect ontological, semantic and epistemic doctrine (see Chapter 3 and Section I of Chapter 6).

(4)   Causal internal realism (especially, or at least, (CIR) of Chapter 6) is a correct view .

(5)   Ontological materialism is correct.

Note finally in connection with (5) that we have sketched a

viable metaphysical position which not only claims that all
there is in the real world is matter (matter understood in its
wide, modern sense) but that we do not need among our basic
ontological entities any realm of abstract entities (constructs,
meanings, universals, propositions, values, or what have you) in
addition to the materialistically understood real world. It is
of course a different matter to speak of concepts, structures,
natural numbers, and so on as quasi-ontological categories. That
is quite all right as long as they are treated only as deriva-
tive and dispensable ontological categories.

CHAPTER 8

SOCIAL ACTION AND SYSTEMS THEORY

I. The Conceptual Nature of Social Action

1. Scientific activity can be looked upon as a social process which consists of joint actions performed by several researchers. In a wide sense this can be understood to be true of not just scientific activity as group activity but also of the foundations and justification of its results. At bottom these ingredients cover the scientific community as a whole, as we shall see in the next chapter.

This chapter aims to develop conceptual equipment for the study of the pragmatics of science, that is, for the understanding of scientists' activities and the thinking which justifies it. A central part of this equipment consists of my theory of social action (see Tuomela, 1984). This theory analyzes various concepts of social action and activity, especially actions performed together intentionally by several agents, as well as so-called we-intentions and mutual beliefs which justify these actions and activities through practical inference. I shall briefly touch on these concepts and tie them to the examination of social processes and changes. In the latter task I shall make use of systems-theoretic tools. But first we need to make a digression into action theory.

Quite generally, by social action I shall here mean joint action carried out by several agents: action in which agents in some suitable way adjust their actions to those of others while pursuing a shared goal, following a mutually acknowledged rule, or engaging in a common practice. Examples of such activities include jointly carrying a heavy table, riding a tandem bicycle, playing tennis, playing Bach's dual violin concerto, greeting, asking questions and answering, and also a group's discussing and solving problems, a nations's declaring a war, etc. (The last two examples deal with actions performed by a collective agent.)

Next, I shall define a pair of concepts. We shall say that

the **result** of my act of opening a window is the opening of the window. The result of an action is an event or state of affairs connected with the action such that performing the action is conceptually impossible without the result (see von Wright, 1971, p. 67). Events and states of affairs which are brought about by an act but which need not be conceptually presupposed by the act are called **consequences** of the act. (The result of an act is therefore regarded as one of its consequences.) Thus if I inadvertently, when opening a window, let in a mosquito, we shall say that the coming in of the mosquito was a consequence (but not a result) of my opening the window. By actions and acts we shall denote **performances**: each action involves a logically built-in result as well as several factual consequences. The result-consequence distinction extends, of course, to social action as well.

The result of a social action can come about either causally or conceptually or, as is commonly the case, in a "mixed" way. Thus we can analyze the carrying of a table by two agents in such a way that its result r, e.g., the moving of the table upstairs, is a causal consequence of the "component actions" of these two agents (of lifting the corner of the table and moving it in a certain way). If the results of these component actions are events r_1 and r_2, we shall say that r_1 and r_2 causally bring about the overall result r.

Let us next think of a social action which consists of two agents' greeting one another by saying, e.g., "Hi" (utterances with the respective results r_i, i=1,2). Now r_1 and r_2 conceptually generate the result r of the action of greeting, for the two occurrences of the expression "Hi" in such circumstances conceptually mean greeting. Thus we could say that, correctly described, the mereological sum r_1+r_2 of the two events r_1 and r_2 just is the result r of the greeting act in these circumstances. (Let us note that we here have to do with the structural analysis of the greeting act - the notion of greeting itself is not thereby analyzed away, nor is its conceptual status in any way denigrated.)

When a nation declares war it does so through its proper representatives. If someone sells his house by using another agent's services this other agent represents him. **Representing** is a third main category of social action generation, on a par with causal (or more widely, factual) and conceptual generation. There is no need here to examine this or the other ways of action generation in more detail (see Chapters 5 and 6 of Tuomela, 1984).

2. Let us assume that John kicks James or walks on James's land. Both of these actions performed by John have something to do with James. But there is nothing performed by John and James together. What is required of joint action? It seems clear that every participant in a joint action must in some sense do something himself (or at least do something via a representative). In addition to this the individual acts of these agents must, one way or other, be connected with each other, for often John and James can do something (such as sing a song) either individually or together. Two separately performed actions do not as such constitute a single social action.

However, social acts are in an important sense based on the component actions performed by the participating agents. In what follows I shall rely on the so-called purposive-causal theory of action I have developed elsewhere (see Tuomela, 1977, 1984). Space does not allow me to go to the details. Therefore I shall confine myself to presenting a summary account of a part of the theory.

In the purposive causal theory of action intentional action is analyzed with the help of a functionally characterized "willing" (or "trying"). The event of an agent's willing to perform an action U is conceptually identical with the event of the agent's **now** effectively intending to bring about by his bodily movements all that the agent sees as necessary for the performance of U. Thus if someone intends to light a lamp (i.e., to perform U) he ends up forming the intention to flip a switch. This intention can then develop into a causally effective intending-episode, the content of which is precisely the agent's flipping the switch with his finger. The last mentioned episode is precisely the agent's "effective" willing episode.

An agent's singular (i.e. individual) action can, from a conceptual point of view, be characterized as a willing-behavior-result -episode, i.e., as a sequence of events $\langle t,...,b,...,r \rangle$ where ordering stands for the causal order between the three events. Here t denotes willing, b stands for the appropriate bodily movement included in the action and produced by t, and r for the result of the action brought about by t and b. According to the purposive causal theory of action the intentionality of a

singular act u of the type U (e.g. opening a window) can now be
described as follows:

(**PC**)   An agent performed an **intentional action** u if and only if
            (1) the agent's behavior (b) purposively generated r;
            (2) there was a conduct plan, K, of the agent which
                involved an end which the agent effectively intended
                to realize then by his bodily behavior (of the type
                believed by him b exemplified) and which he believed
                his behavior will (tend to) bring about or at least
                be conductive to; and
            (3) this effective intending (as a willing, t) and this
                belief together purposively generated the behavior b
                in u.

    I shall comment on the notion of purposive bringing about
in clause (1) of schema (**PC**) later. Clause (2) involves the
technical notion of a conduct plan. A typical conduct plan makes
reference to one or more goals of the agent, i.e., to actions or
states of affairs which he wants or intends to realize. In addition to this it contains reference to beliefs about suitable
means, and presupposes several kinds of background knowledge and
assumptions. The notion of a conduct plan must be understood in
a very wide sense, in order for (**PC**) also to cover expressive
and habitual intentional action.
    Very often (though not always) intentional action is based
on a prior process of practical inference in which the agent has
a goal and beliefs about how it can be achieved and in which he
derives from these that he ought to perform the action. This is
typical of and perhaps also necessary for all rational activity,
scientific activity included. It is not necessary here to go
into the details of the nature of practical inference, nor into
the psychological concepts - such as intention and belief - presupposed by it (see e.g. Tuomela, 1977, Chapter 4).[1]
    Clauses (1) and (3) of the schema (**PC**) are based on the
notion of "purposive" generation. It has to do with a causal or
at least a partially causal relation (for a more exact characterization, see Chapters 9 and 10 in Tuomela, 1977). Purposive
causality is a non-Humean "purpose-preserving" variety of causation which in a sense explicates the notion of a final cause.
Yet purposive causation adds nothing to the ontological furniture of the world. It harbors no ontological commitments but is

rather a notion which expresses the intentional-teleological character of the conceptual scheme. It also in part explicates the special feature of willing that a willing-episode guides and controls behavior - the feature that an agent interferes with the world through his bodily movements and in accordance with the content of his willing. Hence we could also call it cybernetic causality, for control has to do with paying heed to the feedback provided by the world (see Tuomela, 1977, Chapter 9, Tuomela, 1984, Chapter 6, as well as the discussion below).

3. Let us next turn to social actions. What is involved in the fact that agents $A_1, \ldots, A_m$ jointly perform an action of the type U? We noted that each agent must here do something - his own part or component action. Let U_i be A_i's component action. How are the actions U_i to be put together to constitute the social action U?

My claim is that when agents $A_1, \ldots, A_m$ perform action U intentionally, a shared intended goal must somehow be involved. More precisely: in the case of a **fully** intentional singular social action (activity) each agent must possess a relevant group intention. How can this be justified?

Assume that agents $A_1, \ldots, A_m$ sing together a song or play a round of a card game. We can hardly say that they performed the act **fully** intentionally together or as a collective (at least if the group has no special prior organization), if there is a single agent A_i who lacked that group intention. If A_i had performed his component action U_i intentionally but without having the group intention, he would not have acted intentionally together with the others. Therefore the agents $A_1, \ldots, A_m$ would not have acted intentionally as a group, and their social action would not have been fully intentional.

Another argument for the presence of a we-intention is the following. Assume that both A and B give C a dose of poison sufficient to kill C. Also assume that A and B act without knowing of one another, and that both intend to kill C by poisoning. Can we then say that A and B killed C together intentionally? Clearly we cannot - rather, each one of them performed the action of killing (or, perhaps better, participating in killing) alone and separately. Both A and B have the same intention, and they also acted on it. But joint action would have required that they shared awareness of one another's (and their

own) intention. This leads us to the requirement of the presence
of a we-intention, as can be seen from the analysis (**WI**) of we-
intentions to be presented below.

Let us still look at the issue from the point of view of
the action of the group. We shall again ask: Why isn't the mere
presence of the same intention in all the members of the collective sufficient? Let us suppose that the joint action U fails.
For instance, one agent may fail in his performance of a component action. At least ideally the other members of the group
will give a hand (or, perhaps, exert pressure, etc., as the case
may be) and do their best to secure the performance of the
action. This again renders support to the view that all the
members of the group must believe (or know) that each member of
the group intends to perform U (or at least his share of it). In
fact this requirement concerns the agents' mutual beliefs. At
least ideally, in its strongest form, the **mutual** belief of a
group of agents consists in that everyone believes that each
member intends to perform U, and that everyone believes that
everyone believes that each member intends to perform U, etc.,
in principle ad infinitum. Iteration of the expression ("operator") 'everyone believes that...' justifies lower-order beliefs
in a quite normal way (see Tuomela, 1984, Chapter 7, where also
weaker types of mutual belief are discussed).

Whether or not the agents $A_1,...,A_m$ act intentionally together or separately depends, then, essentially on whether they
act on a "mere" I-intention or a we-intention - some relevant
intention is needed in any case. Let us note that the intentional performance ⟩of U does not necessarily presuppose an
intention to perform just U. (We can also define weaker notions
of intentional social action. For these, see Tuomela, 1984,
Chapter 5.)

II. We-intentions and Social Action

1.   How are we to characterize, in more precise terms, the we-
intentions of individuals? Let U be some social action (viz.,
type of action). Apparently we must require that an agent intends either to perform U (e.g., to sing a song or to pick a
bucketful of berries) or his part or share of U (as in the
social activity of questioning and answering). An agent's having
a we-intention in any case involves his intention to do some-

thing appropriate. Another element to be found in a we-intention
is precisely a mutual belief, as already said.

I shall next present my definition of this motivational
notion of we-intending, without further explanation or justification (cf. Tuomela, 1984, Chapter 2, Tuomela and Miller, 1985).
It is related to a group or collective A which comprises the
agents A_i, i=1,...,m:

(WI)   A member A_i of a group G **we-intends** to perform U if and
        only if
        (1) A_i intends to do his part of U, given that he be-
            lieves that every member of G (or at least suffi-
            ciently many of them, as required for the performance
            of X) will do his part (their parts) of U;
        (2) A_i believes that every member of G (or at least
            sufficiently many of them, as required for the per-
            formance of X) will do his part (their parts) of U;
        (3) there is a mutual belief in the group G to the effect
            that (1) and (2).

This definition includes two important "social" features.
First, A_i's intention to do his part of U, in clause (1), is
conditional on the actions of the other agents in the group - A_i
would not alone set out to perform his part of U. Secondly, it
can be demonstrated, on the basis of (1) - (3), that A_i intends
(unconditionally) to do his part of U (partly) **because** he believes that all members of G will do their parts of U. Note in
addition that on the basis of (3) we-intentions are intersub-
jective. (Definition **(WI)** and questions closely related to it
have been discussed more closely in Tuomela, 1984, Chapter 2 as
well as Tuomela and Miller, 1985 - but the above should suffice
here.)

2. On the basis of what has just been said we can now describe
(fully) intentional social action by help of a short formula
which generalizes the one-agent formula (**PC**) to the case of
several agents. Let us assume that agents $A_1,...,A_m$ form a group
in some loose sense which does not presuppose any specific organization (see Tuomela, 1984, Chapter 8). If the agents $A_1,...,$
A_m have together performed a singular action u (of the type U),
where their own singular component actions are u_i (of the type

U_i, i=1,...,m), then our purposive-causal theory of action gives its intentionality the following characterization in which U is a kind of conjunction of the component action types U_i:

(**PCS**) The agents A_1,...,A_m (jointly) performed an **intentional social action** u if and only if
  (1) the results r_i, i=1,...,m, of the agents' component action tokens u_i together purposively generated the total result r of u;
  (2) there were conduct plans, say K_1,..,K_m, of A_1,...,A_m, respectively, which involved an end action the agents effectively we-intended to realize then by their bodily behaviors (of the types they took the b_is to exemplify) such that they believed their respective behaviors will (tend to) bring it about or at least be conducive to it;
  (3) the agents' effective we-intendings (as we-willings) and the beliefs referred to in clause (2) together purposively generated their behaviors in the u_is, and intermediately, the results r_i; and
  (4) each agent A_i performed his component action u_i intentionally.

(**PCS**) presupposes that u is a singular social action. We shall enquire into this matter more closely below. Clauses (1) and (3) speak of purposive generation, a notion which will not be discussed further here (cf. (**PC**) above and Tuomela, 1984, Chapter 6). The we-willings in clause (3) can be understood as we-intendings to perform U **now**. The rest of (**PCS**) ought to be understandable on the basis of what was said above in the schema (**PC**).

What, then, is the notion of singular action on which (**PCS**) relies? It refers to singular action in a very general sense, and, furthermore, one which does not depend on any specific type of action. A singular social action is now considered to include the result event r (as is already assumed in (**PCS**)). The results of the component actions of the participating agents are required to generate it (cf. (**PCS**)). A singular social action also includes the behavior events involved in the component actions of the agents (such as are the events b_i in schema (**PCS**)). In addition, in encompasses as components the kind of "social" mental events which correspond to the we-intentions of (**PCS**) in

this general case. Each such social mental event is assumed here
to instantiate some so-called "we-attitude", viz. social propositional attitude, or at least to be a consequence of such an
instantiating event.

This notion of a singular social action (activity) can be
shown to be very comprehensive, and it includes, e.g., all non-
intentional actions. However, I cannot justify this claim more
fully here (see Tuomela, 1984, Chapter 5). My thesis, however,
is that it covers everything that a group of agents can jointly
do, in the broadest sense. On the basis of such general examination we can then classify social actions in many ways, as I
have in fact done elsewhere.

3. Let us now, as an application of the above, consider a
group of scientists $A_1,...,A_m$ who intend to carry out a research
programme. They can, for instance, be medical doctors interested
in the effect of interferon on lung cancer, or social scientists
who have set as their task the construction of a theory of
social change. Their intentions can of course be based on widely
different motives, but these are of no concern here. The important thing is that a scientific community, and consequently its
members, can not form just any kind of intentions when acting in
accordance with the standards of the method of science. The
practical inferences which justify scientific activity and
scientific action-intentions must fulfil certain rationality
conditions. These conditions are based on the fact that science
should be at least objective, critical, autonomous, self-corrective and progressive (see Chapter 10 below). I shall not
here investigate how these objectives can be connected with
acceptable practical inferences in science, but it appears obvious that they may exert influence on the scientific practice.
Thus, for instance, neither an individual scientist nor a scientific community should form a we-intention to give a purely ad
hoc explanation or to test a theory by means of a test which
fails to fulfil the requirement of repeatability of experimental
results.

I have elsewhere studied various models of practical inference which are associated with social action (see Tuomela,
1984, Chapter 7). In what follows I shall present one of them,
without further explanation. Let us examine a social action U
which can for instance be an act of writing a joint scientific

paper or of together carrying out an experiment. We can think
that an agent A_i, i=1,...,m, reasons in accordance with the
following schema:

(PR_i)($P1_i$) I intend to perform my part of action U, provided
that all agents A_j, j = 1,...,m, j≠i, will perform
their parts of U.

($P2_i$) I believe that unless we, i.e. A_1,...,A_m, each perform our parts of U, we cannot together perform
action U.

($P3_i$) I believe that A_j, j = 1,...,m, j≠i, will perform his
part of U.

($P4_i$) I believe that A_j, j = 1,...,m, will perform U_j as
his part of U.

($P5_i$) I believe that normal circumstances for the performance of U obtain in this situation.

(($P6_i$) We, i.e. A_1,...,A_m, mutually believe that (($P1_j$)&
($P2_j$)&($P3_j$)&($P4_j$)&($P5_j$) and therefore that (C_j)), for
each value of j = 1,...,m.)

(C_i) I will perform action U_i.

It can be demonstrated that schema (PR_i) is a conceptually
valid one (see Tuomela, 1984, Chapter 7). Let us note here that
premiss ($P6_i$), which is included in the background knowledge
presupposed, implies that for any two members A_i and A_j of the
group there is a loop belief 'A_i believes that A_j believes that
A_i...', which can be iterated.

Schema (PR_i) implies that A_i **we-intends** to perform action
U. This follows from the definition (**WI**) and from the above
premisses ($P1_i$), ($P3_i$) and ($P6_i$). Thus we could have used, instead of ($P1_i$), the premiss

($P1_i'$) We will perform U.

The schema (PR_i) covers in principle a wide variety but not all,
of the activities based on practical inference - after all, it
is a rather simple one. It is to be observed that all social activities need not be based on practical inference, not even all
those that are intentional social actions. Nevertheless all
types of social actions **can** be performed by grounding them on
some relevant social practical inference.

Action U in the above schema need not confine to 'concrete'

actions, such as the performing of an experiment. It can be interpreted to cover actions which are performed for the maximization of epistemic utilities (e.g. explanatory power, truth) or, say, replacing one theory by another. We shall keep these wider possibilities in mind in later chapters. In Chapter 9 I shall also examine another type of scheme of inference relevant for scientific practice, one which in a sense can also justify premise $(P1_i')$ of schema (PR_i) for some special cases.

III. Joint Action and Systems Theory

1. When we examine scientific progress from the point of view of social practice - and it is our intention to do so - and therefore from a pragmatic point of view, we come to grips with processes of social action and social change. More particularly, we need to be able to tie social action to some kind of possible social laws of development which depict scientific progress. To this end I shall in the sequel employ a kind of systems-theoretic approach. Since this in its turn requires a partial mathematization of the notion of activity (action), I shall start with this task.

The main idea is to look upon actions (types of actions) as representable by mathematical structures of a sort. Because we are examining activities as performances and achievements it comes as natural to consider an agent's mental events, such as intendings, as independent variables. In the same vein the appropriate bodily movements form an independent variable (despite the fact that they are brought about by the prior mental events). The results and consequences on the other hand are thought to be dependent variables.

Recall that a singular intentional action (of kind U) can be represented by a sequence ⟨t,...,b,...,r⟩ in which t is a willing episode, b an appropriate behavior event, and r an appropriate result. In non-intentional action willing is replaced by what I have called a volition (cf. Tuomela, 1977, Chapter 10). However, here we wish to restrict our attention to intentional actions, for the actions we want to study can well be taken to fall into this category. In most cases they are also based on practical inference. From now on we shall accommodate actions which are not based on inference as special cases in which willing comes about as a result of "empty" practical inference.

We shall now examine an action type U which can be arbitrarily complex (for this, see Tuomela, 1984, Chapter 5). Our proposal is that U can be represented by a function (f_U) as follows:

(1) f_U: P × B × S → Z

Here P is a class of singular, causally efficacious practical inferences. We shall assume that as consequences of such inferences we have willings to perform U. B is the class of appropriate behavior events. Members of these classes are brought about by the corresponding willings. S is the class of circumstances and situations which specifically includes the internal and external normal conditions required by the intentional performance of U. (Incorporating S is central, but I cannot here give a more fine-grained "situation analysis".)

I have talked as if all token occurrences of U were intentional and based on practical inference. If we also want to include non-intentional token occurrences of U we can proceed as follows: we include in the set P also volitions and drop the requirement that they always be conclusions of practical inferences. In addition to this, we do not require that the behavior events be **purposively** caused.

Z is the set of all such (physically possible) singular total consequences (i.e, of complex result-consequence events) as are brought about by the agent's attempt to perform U; in particular, Z includes all the results of U. In some applications Z can be interpreted to include only the results of U, but we shall not adopt this interpretation for the general case (about the definition of the set Z, and its dependence on the conceptual scheme, see Tuomela, 1984, Chapter 12).

According to formula (1) the action function f_U then involves the following. An intentional singular one-agent action based on inference is an ordered quadruple ⟨p,b,s,w⟩ in which p∈P is a singular, causally efficacious process of practical inference, the conclusion of which is a willing (cf. (**PR$_i$**)); b∈B is an appropriate behavior event; s∈S is a singular, possibly complex situation or circumstance; and w∈Z is a singular total consequence brought purposively about by the joint occurrence of p and **b**.

2. Next, I shall define some general notions of a system which are connected with our discussions in Chapter 2, and which are needed for further developments. A system is a structure which consists of a set of parts of some entity x, of a set of entities representing the environment of x, and of relations defined on these two sets (see Bunge, 1979). Let us give a more precise definition for these elements by assuming there to be some set O, which can be a set of real entities or, alternatively, a set of (total) states of real entities. We can say that, relative to an environment, a system involves a set of entities (or states) which interact with one another, and which are parts of some entity x. These parts form the composition of x. In other words, we can define the **composition** K of x as the set

$$K(x) = \{y \in O \mid y [x\}$$

where '[' stands for a part-relation, taken here to be a well-enough understood concept also when the elements y and x are (total) states or processes. We can now relativize a composition to some set $B \subseteq O$ as follows:

$$K_B(x) = K(x) \cap B = \{y \in B \mid y [x\}$$

Having done this we define the notion of a (relative) environment starting with the assumption that we can avail ourselves of an intelligible notion of influence ($\Rightarrow$): 'a$\Rightarrow$b' is to mean 'a influences b'. We shall define the notion of an **environment** E_B relative to the set B as follows:

$$E_B(x) = \{y \in O \mid \sim (y \in K_B(x)) \& (Ez)(z \in K_B(x) \& (y \Rightarrow z \vee z \Rightarrow y))\}.$$

In other words the B-environment of x comprises all entities (or states) y which do not belong to the composition of x and which interact (in one direction or other) with some element in the composition of x.

We can then define the **structure** of x as consisting of all those relations which obtain between either a) the elements of $K_B(x)$ or b) between the elements of $K_B(x)$ and $E_B(x)$. Let this set of relations be $R_B(x)$.

We are now in a position to define that a system in relation to an entity x and to a set B is a structure $\sigma_B(x) = \langle K_B(x), E_B(x), R_B(x) \rangle$, in which at least the sets $K_B(x)$ and $R_B(x)$ are non-empty and $K_B(x)$ includes at least two elements. $E_B(x)$ may be empty - we then call the system closed.

Alternatively, we could say that an entity x is a system if it can be described by means of the above triple. The set B can be assumed to be the set of the atomic parts of x. It is natural to think that, e.g., a social group which comprises a set of agents, B, forms a system. (x can here be thought to be a total- ity of entities formed by all its agents - no matter how exactly such a totality is understood.)

A more thorough characterization of a system requires that we also lay bare the history, and above all, the laws of the system. The specification of the laws of development is es- pecially true of **dynamic** systems. In those cases we have reason to relativize the notions $K_B(x)$, $E_B(x)$ and $R_B(x)$ to time.

Let it be noted here that the composition relation allows us to define "ontological" levels. x and $y \in K_B(x)$ are (or can be) entities at different ontological levels. On the face of it it would be better to say that in a real system at least the el- ements of $K_B(x)$ are real in the sense of being capable of exert- ing causal influence. In contradistinction to this x need not be real in this concrete sense.

3. We shall now apply the above generally defined notions to the study of action processes. The application goes as follows (un- fortunately I am forced to trouble my readers with a change of notation - for reasons of presentation). Corresponding to the entity **x** in the above definition there will be the agent (**A**). We shall assume that the agent's states (episodic ones included) can be thought to be parts of the agent in the sense of the [- relation. (This is not such a demanding assumption, for [was earlier left as a kind of primitive notion.) Hence K(A) will consist of all those (total) states of the agent we want to incorporate into our examination; in other words, we shall rela- tivize it to some set B. We now adopt a new set of symbols and define $X_A =_{df} K_B(A)$; i.e., X_A is the - widely construed - set of (total) states of the agent A. Furthermore, we define $\bar{X}_A =_{df}$ $E_B(A)$. We do not below need an exact counterpart to $R_B(A)$ but we shall introduce a T_A to relate X_A and $\bar{X}_A$ and in fact require that T_A be a transition-function for states in a sense to be specified later. Thus the triple $\langle X_A, \bar{X}_A, T_A \rangle$ which corresponds to an agent A can be looked upon as a system in the sense given to

this notion above. Alternatively, agent A can itself be called a system, as we in fact will do later on.

The total state of a system A, be it x_A, and the total state ($\bar{x}_A$) of its environment, together define the state (x) of the world to be studied. Consequently we adopt the following definition:

(2) $x = \langle x_A, \bar{x}_A \rangle; \quad X = X_A \times \bar{X}_A.$

We shall now investigate the actions of the system A. We shall accept the rather natural idea that an action of A, e.g., $f_{U,A}$, must be a function of the states of the system, i.e. the domain of the function $f_{U,A}$ must be X_A. But what is the range of this function? We shall generalize our earlier proposal (1) and propose the following:

(3) $f_{U,A}: X_A \to X$

In other words, the range of the function $f_{U,A}$ is defined as the set X of all (physically possible) states of the world. Yet, (3) can be tied closely to our previous definition (1). For according to (2) $X = X_A \times \bar{X}_A$, and we now use in our application the definition

(4) $X_A = P \times B \times S; \quad Z = X.$

It follows that we can in fact regard formula (3) as a technical generalization of formula (1). Every triple $\langle p,b,s \rangle$ of the Cartesian product $P \times B \times S$ is mapped into some quadruple $\langle p',b',s',w \rangle$ in X. Since the technical examinations will run more smoothly using definition (3) than definition (1), we shall use (3).

The range of the action function $f_{U,A}$ defined by (3) is the set X ($= X_A \times \bar{X}_A$). To be able to discern better the effects of such actions on the states of the system A itself, that is, on X_A, we define a projection function

(5) $P: X \to X_A$

which simply drops from every quadruple of the type $\langle p,b,s,w \rangle \in X$ its last element w, so that $P(\langle p,b,s,w \rangle) = \langle p,b,s \rangle \in X_A$, where $p \in P$, $b \in B$, $s \in S$, $w \in \bar{X}_A$ and $X_A = P \times B \times S$.

We now move on to study the dynamics of the system A.[2] (A
reader who is not interested in technical details may read   the
following   pages   in a cursory manner and confine   to   the   main
conceptual   import   of the "growth function" (**18**) to be   derived
later.) We shall begin by relativizing function $f_{U,A}$ to time.
Let T be a set of instants (or, alternatively, intervals) of
time. We then investigate actions as a kind of time segments of
the following kind:

(**6**) $f_{U,A}^{t\tau}(x_A(\tau)) = x(t), \ t \geq \tau,$

where $t, \tau \in T$, and where $x(t) \in X$ represents the total consequence
(a singular state or event) at (or up to) time t of the   occurrence of the state $x_A \in X_A$ at time τ.

We can now examine what effects the performance by A of an
action U at τ has on A itself at t, $t \geq \tau$, and how we can in this
way find a state transition function for the system. Let $x_A(\tau) \in$
X_A be A's total state at τ. According to formula (**6**), performance of U leads to the state of the world $x(t) \in X$ at t. We then
apply to this, with appropriate relativization to time, the
projection function P (formula (**5**)). P will then give $x_A(t) \in X_A$
as the projection of $x(t)$. In other words

(**7**) $x_A(t) = P(f_{U,A}^{t\tau}(x_A(\tau))).$

We can now define the transition function for A:

(**8**) $T_A^{t\tau} = P \circ f_{U,A}^{t\tau},$

and hence

(**9**) $x_A(t) = T_A^{t\tau}(x_A(\tau)).$

Formula (**9**) shows that the just defined state transition   function or "mode of action" $T_A^{t\tau}$ defines the system A as a   dynamic
input-output -system, with $x_A(\tau)$ as its inputs and $x_A(t)$ as its
outputs, with $t \geq \tau$. Recall, however, that (**9**) depends on the
action-consequences of the system in the world and thus takes
into account the impact of the system on the world and of the
world on the system.

In terms of the interpretation intended here the mental
states (and episodes) of the agent A, i.e, the elements of the

set P, and the behavior episodes of the set B bring about, in
favorable circumstances (S), new mental states, behavior episodes and states by means of feedback interaction with the
world.

Note that we have here only dealt with deterministic (and
causally understood) relations in formulas (1), (3) and (6). By
employing random variables we could also study probabilistic
connections between the sets P×B×S and Z, and hence also the
state transition function T_A would be probabilistic.

4. One of the central aims in this chapter is to connect actions
performed by several agents, i.e, social actions, with social
system-laws which are of the form (9) or slightly more complex,
as we shall soon see. For these purposes we shall study agents
$A_1,\ldots,A_m$, m>1, which can all be represented as dynamic systems
in the above way and which are members of the collective G, in
our earlier terminology. Assuming that these systems are distinct (set-theoretically disjoint), they form a collective agent
A whose total state X_A, action function $f_{U,A}$ and state change
function can be expressed by means of the values of the corresponding characteristics of the agents $A_1,\ldots,A_m$ as follows:

(10) x_A = $\langle x_{A1},\ldots,x_{Am}\rangle \in X_{A1} \times \ldots \times X_{Am} = X_A$
         $f_{U,A}$ = $\langle f_{U1,A1},\ldots,f_{Um,Am}\rangle$
         T_A = $\langle T_{A1},\ldots,T_{Am}\rangle$.

In (10) it is possible to think that $X_{Ai}=X_{Aj}$ for all i,j
because the elements in these sets are (possible) mental states
any agent can be in. Analogously, I see no harm in assuming for
our purposes that $\bar{X}_{Ai}=\bar{X}_{Aj}$ for all i,j, viz., that the environ-
ments of these agents can be taken to coincide in our applications, for $\bar{X}_A$ is concerned with all possible environmental situations. (This does not exclude the consequences of your particular actions being conditions of my particular action, for instance.)

Disregarding time considerations, we may now pose the problem of how the consequences of the action $f_{U,A}$ depend on the
actions of the individual agents, and especially on the consequences of these actions. Let us assume here that $f_{U,A}$ stands
specifically for a **joint** action of the agents $A_1,\ldots,A_m$, and
therefore something which they do jointly. I have previously

studied this problem in detail with respect to the results of
the actions, and also presented an exact solution (see Tuomela,
1984, Chapter 6). There I used the theory of automata to present
the case in which the singular actions $u_1,\ldots,u_m$ (of types
$U_1,\ldots,U_m$) performed by agents $A_1,\ldots,A_m$ generate a social act
of type U, when (in a sense) $U=U_1\&\ldots\&U_m$. Quite generally, the
problem is that not all possible ordered sequences of consequences $\langle x_1,\ldots,x_m\rangle$ qualify as standing for consequences of
social action, i.e, all combinations do not represent joint acting. Therefore (10) does not suffice for our purposes. In our
present framework (and omitting reference to time) we can say
that we are looking for a "jointness-function" j_A in the expression

(11) $f_{U,A} = j_A(f_{U1,A1},\ldots,f_{Um,Am})$

or, more concisely, $f_A=j_A(f_{A1},\ldots,f_{Am})$. I shall not here repeat
the results of my automata-theoretic developments, but rather
confine to point to some central structural kinds of cooperation. As I have shown in the mentioned studies, it is sufficient
that we in this connection examine the results and consequences
of the actions and activities. Hence I shall present the following central possibilities as examples by means of consequence
states and events:

(12) $r \gg r_1{}^{\cdot}\ldots{}^{\cdot}r_m$

(13) $r \gg r_1+\ldots+r_m$

(14) $r \gg r_1\ldots r_m.$

Here r_i is the total consequence brought about by the realization of the action function $f_{Ui,Ai}$ or, if so wanted, some part
of it (e.g., the result of the action in the sense of subsection
1); r is the corresponding total consequence of the collective
action or a suitable part thereof (see Tuomela, 1984, Chapter
5). The symbol $\gg$ for the mereological part-relationship is
defined for singular events or states, + stands for the sum
(which corresponds to a disjunction) defined for them (in the
sense of the so-called calculus of individuals) and $\cdot$ in its
turn stands for their product (corresponding to a conjunction);
$r_1\ldots r_m$ is an ordered sequence of individual events.

Of the possibilities laid down (12) seems to suit well such cases as carrying together a table or, to take an example from science, writing a joint scientific paper. Formula (13) employs sums instead of products. The proposal appears to fit e.g. problem-solving in a group: a collective A has solved a problem if one of its members has solved it. Proposal (14) makes use of sequences, and it seems to capture actions such as querying and answering. Querying and answering as a social activity presuppose, namely, a sequence of the result of a question and that of an answer among its total consequence (cf. also scientific explanation, Chapter 9).

Let us now connect the case of several agents up with the one-agent case also with respect to laws of change. Corresponding to formula (6) we get the formula

(15) $f_{U,A}^{t\tau}(x_A(\tau)) = x(t)$, where $t \geq \tau$, $x_A \in X_{A1} \times \ldots \times X_{Am} = X_A$
  and $x(t) \in X_1 \times \ldots \times X_m = X$

According to formula (15) all the agents perform their component actions at (or during the time-interval) τ. A more general case is obtained when the agents $A_1, \ldots, A_m$ perform their parts at (possibly) different times $\tau_1, \ldots, \tau_m$. However, we cannot investigate this possibility any further here.

In the case of joint action our agents' single action must be combined in the right way to make up a total action qualifying as a joint action. This was supposed to be taken care of by our **jointness-function** j_A (formula (11)). In order for the action $f_{U,A}$ to generate a state transition function for the system A we must in addition require that $f_{U,A}^{t\tau}(x_A(\tau))$ determines the state $x_A(t)$ for all values of the time parameters t, τ such that $t \geq \tau$. But this requirement is immediately seen to be satisfied within our set-up because of the restrictions that the jointness function j_A imposes. To express this in a mathematically neat way we will consider the situation from the perspective of X_A (= $X_{A1} \times \ldots \times X_{Am}$) rather than X (= $X_1 \times \ldots \times X_m$). Let us thus consider the analogue of the projection function defined by (5) for the present case, understood to lop off the last element of each component of the total consequence x(t), viz.

(16) $P(x(t)) = x_A(t)$, where $x(t) \in X$, $x_A(t) \in X_A$.

As (16) shows P is now defined for a collective agent

rather than a single agent.   Here x(t) deals with any points   -
not just values of j_A. The problem that we now face is to select
those points $x_A(t)$ in X_A which serve to determine a state-transition function for A.   Only such points qualify as are obtained
from the values of the j_A-function. Indeed, these points must be
projections   into X_A of the values of j_A.   Rather   than   writing
this out explicitly in this way it is neater to define a counterpart jointness function J_A for the set X_A such that the
values of the function $J_A(x_A(t))$ are states of X_A determined by
$f_{U,A}$ if and only if $x_A(t)=P(x(t))$ such that $x(t)$ is a value of
the jointness function j_A.[3] Thus recalling that then x(t)=
$\langle x_1(t),\ldots,x_m(t)\rangle$ where $x_i(t)=f_{Ui,Ai}^{t\tau}(x_{Ai}(\tau))$, we get

(17) $x_A(t) = J_A(P(f_{A1}^{t\tau}(x_{A1}(\tau)),\ldots,f_{Am}^{t\tau}(x_{Am}(\tau))))$, where $t \geq \tau$.

But as $f_A^{t\tau}(x_A(\tau))=\langle f_{A1}^{t\tau}(x_{A1}(\tau)),\ldots,f_{Am}^{t\tau}(x_{Am}(\tau))\rangle$ we can write the
obtained state-transition function in a compact form as

(18) $x_A(t) = T_A^{t\tau}(x_A(\tau))$, where $t \geq \tau$ and $T_A^{t\tau}=J_A \circ P \circ f_A^{t\tau}$.

Written in this way, **(18)** closely resembles the state-
transition function **(9)** for the single agent case. However,
there is a central difference, for in **(9)** $T_A^{t\tau}=P \circ f_A^{t\tau}$ whereas
above in addition the function J_A has been added to the composi-
tion.   The function J_A - which was defined to the counterpart of
the jointness function j_A - is a coupling function or operator
in the cybernetic sense. It couples the single agents' actions
in the right way to yield a joint action and a state $x_A(t)$ of
the system A determined by its previous state $x_A(\tau)$. So we may
say that from a conceptual point of view the state-transition
function $T_A^{t\tau}$ is composed of two central parts, viz. J_A and
$P \circ f_A^{t\tau}$. Defining $V_A^{t\tau}=P \circ f_A^{t\tau}$, we may call V_A the transition part
and J_A the coupling part. Viewed from another perspective, we
may regard **(18)** as a candidate for a social law (or schema of a
social law).   It is a systems-theoretic process generalization
expressed in a mathematical language.
All in all, $T_A^{t\tau}$ gives, in the case of social action, a partial explicate for what is meant by the self-regulating nature
of a collective. For it incorporates not just the requirement of
cooperation but also the requirement that the agents can collec-
tively regulate their cooperation (in other words, have an
impact on the systems input states). Such "we-regulation" or

"we-control" is not needed in the one-agent case, but of course it is crucial for the case of several agents. We can think that it is a consequence of the appropriate we-intentions of the agents $A_1,\ldots,A_m$, and the practical inferences associated with these intentions (cf. formulas (PCS) and (PR$_i$) of section II). The state transition or growth function $T_A^{t\tau}$ takes into account the feedback from the agents' previous actions, and we can think that it in a similar coordinating way adapts these practical inferences (i.e., effects appropriate selection in the set P), on which action is founded. (Function $T_A^{t\tau}$ can in principle be randomized, but I shall not go into this here; see above.)

5. In the next chapter we shall inquire into scientific development, in part from the point of view of systems theory. To this effect we must define a few additional systems-theoretic concepts.

In section II I restricted the treatment to one type of social activity (or action), which, to wit, can be thought to be arbitrarily complex (in the precise sense of Tuomela, 1984, Chapter 5). But doesn't an examination of scientific progress require simultaneous treatment of several types of social action? Apparently the answer to this question is positive, if by types of action we mean ordinary ways of speaking about acting, or labels of action types to be found in ordinary language. But when we examine types of action in the above sense (cf. also what was presented in (PCS)), this is far from clear. For U can, e.g., represent the replacement of a scientific theory by another - such replacement can be regarded as a social action (cf. schema (2) in Chapter 9). Our solution in this and the following chapter will be that although we wanted, on the basis of criteria obtained from social practice, several types of social action, we nevertheless assume that they have been united to be treated under one function of the form (15). We shall restrict our examination to this case which, as will be seen in the next chapter, nevertheless is not quite unrealistic.

Let us examine formula (18) and the growth function $T_A^{t\tau}$. Let $X_{t0} = \{x_A(\tau)|\tau < t_0\}$ be the set of the states of a system A prior to the instant of time t_0. By letting the time variable vary we get the following sequence of sets of states

(19) $\quad X_{t0} \subseteq X_{t1} \subseteq X_{t2} \subseteq \ldots$, where $t_0 < t_1 < t_2 \ldots$

As far as self-regulating and creative systems are concerned there are reasons to think that the system never returns to a state already passed, and that the sets X_{ti}, $i=0,1,2,\ldots$, all are non-identical with one another. This of course depends on how time has been sliced and how exactly the states of the system have been defined.

Next we shall investigate a system A which obeys the change function $T_A^{t\tau}$. We shall examine the behavior of A in the light of the function, letting t grow without limit. On the basis of this examination we can in fact classify cybernetic systems (a cybernetic system is a system with at least one feedback loop). We can, namely, draw within cybernetic systems a division between goal-directed and disintegrative systems as follows (cf. Aulin, 1982, Chapter 2). A system is **goal-directed** if it is directed towards a goal in such a way that

(20) $x_A(t) = T_A^{t\tau}(x_A(\tau)) \to g_\tau(t)$, when $t\to\infty$, where $x_A(\tau)\in D_\tau \subset X_A$.

Here $g_\tau(t)$ is a function which in some special cases can be a constant. The set D_τ is called the **domain of stability** of the system A at τ. D_τ sets the limits within which external disturbing factors up to time τ can have changed the state $x_A(\tau)$ in a way which still makes it possible that, according to the function $T_A^{t\tau}$, the system will still reach its goal $g_\tau(t)$. In the case in which $g_\tau(t)$ is either a constant or a periodic function (i.e, $g_\tau(t+c)=g_\tau(t)$ for all values of t), we shall call the system a **self-regulating** goal-directed system. Otherwise, i.e., when $g_\tau(t)$ is neither a constant nor a periodic function, the system is called **self-directing.** (It may occasionally turn out to be realistic to replace the condition $t\to\infty$ in (20) by the condition $t \to t'$, where t' is a sufficiently large number.)

All goal-directed systems allow the partition of D_τ into sets $D_{\tau,1},\ldots,D_{\tau,n}$ in such a way that these partitions are, when the growth function is applied, connected with different goals $g_{\tau,i}(t)$, $i=1,\ldots,n$. On the other hand we can still note that if the area of stability vanishes completely, the system is not goal-directed but **disintegrative,** for, due to external disturbing factors, the system is not in an area from which the growth function $T_A^{t\tau}$ could take it to a goal.

6. Are scientific communities goal-directed systems in the sense of cybernetics? The question is a vexing one. No doubt the majority of philosophers agree that scientific communities can be studied as goal-directed agents. However, I suspect that relatively few of them would be willing to consider them to be goal-directed in the sense of definition (20) above, if it is required that the function $g_\tau(t)$ is a constant (or a periodic function). But goal-directedness in the sense of genuine self-regulation has a great deal of appeal, for it would enable progressive, ever-growing science.

The systems theoretical assumptions concerning scientific (and other) communities made in this chapter do not enable us to give an a priori answer to the question if actual scientific communities are self-regulating in the sense defined there. This is because finding the factually correct growth function $T_A^{t\tau}$ (cf. formula (18)) is of course a matter of empirical investigation. On the other hand it is clear that when such a function has been specified we can use mathematical methods to examine if any given system is self-regulating. If we can, furthermore, connect to the function $T_A^{t\tau}$ such epistemic matters as the growth of the explanatory power and truthlikeness of the theories accepted by a community at different times, we could, by studying that function, ideally say much about where science is heading. Thus if $T_A^{t\tau}$ is an analytical function we can study its mathematical behavior by help of its Taylor-series and specify under what conditions it represents goal-directedness, self-regulation, and so on. I shall not here go into the technical details. Rather, I shall comment on the general conceptual relationships between practical inference and the growth function.

Recall that our function $T_A^{t\tau}$ is tied to a community or group A and a social act U. We can think that the community A consists of the scientists engaged in research on a realm of the world, say R. We have assumed above that action U is, in the situations examined, performed intentionally in accordance with a practical inference in such a way that the inference suitably incorporates the we-intention of the members of A (cf. set P in formula (1)).

Our growth function $T_A^{t\tau}$ deals with one agent collective A and one, possibly complex, action U. This is to idealize somewhat at least. Above we already discussed the limitation to one action U (e.g. in case of substituting one theory for another).

As to the assumption of one agent collective, we can say, to begin with, that we can in principle allow the members of A to be groups of agents (and not merely individual agents). Consequently the members of A can be, say, research groups all of which are engaged in research on the portion R of the world. They can compete with each other in many ways, but we shall assume that they nevertheless cooperate on a very general level, e.g., in attempts to develop a basic theory for R (for this, see schema (2) to be presented in Chapter 9).

To put it quite generally, we ought to allow that the area R is investigated by several rival research groups, say $A_1, \ldots, A_n$, which ground their work in quite different research programmes and which therefore resort to correspondingly different practical inferences. This assumption would require us to study growth of knowledge from a group-dynamical sociopsychological point of view in which different groups were represented by different growth functions. We cannot undertake such a complex task here, nor is it in fact necessary for our present purposes. Note, however, the following. Even if we had an initial stage in which we would have to examine groups $A_1, \ldots, A_n$ with radically different research programmes, these groups might be describable, with sufficiently great values for t, by the same growth function. The reason is that they might perform a joint action U (e.g. that of replacing a theory T_i by another one T_{i+1} for some i) although they might have clearly distinct grounds for performing that action.

We shall nevertheless assume in what follows that our object of study is a collective A which is monolithic in the sense that it draws on the same grounds for action. Thus the members of A generally ground (or may ground) their experimental and theoretical activity on schema (PR_i) of Section II, above, or on some similar, possibly much more complex, schema. Theory-replacement, in its turn, is performed in accordance with a schema for we-intending (see especially schema (2) of Chapter 9), at least for high enough values for t. The general validity of the assumption of monolithicity may be hard to defend, but it is not always totally unrealistic.

7. To end this chapter, we shall consider as an application a situation in which there is a collective A, which consists of scientists $A_1, \ldots, A_m$. A investigates some region, R, of the

world and is in causal interaction with it. We shall now examine
system S which consists of the collective (system) **A** and the
region (system) **R**. Using basically the symbolism (but without
the time index) employed earlier in subsection 4 (see the discussion related to (**18**)) we can define (cf. Aulin, 1982, pp. 95
- 97):

$$(21) \quad x_S = \langle x_A, x_R \rangle \in X_A \times X_R = X_S$$

and assuming that the environment of S is constant,

$$(22) \quad \begin{aligned} x_R' &= J_R(y_R, y_A), \quad y_R = V_R(x_R); \quad x_R' = J_R(V_R(x_R), V_A(x_A)) \\ x_A' &= J_A(y_A, y_R), \quad y_A = V_A(x_A); \quad x_A' = J_A(V_A(x_A), V_R(x_R)) \end{aligned}$$

where the primed values x_R' and x_A' refer to the application results of the coupling operators J_R and J_A applied to the outputs
y_A and y_R.

For the sake of simplicity we shall now assume that the
inputs of both the system **A** and the system R can be "separated"
with respect to the coupling operation, viz. that the internal
couplings of A and R are unaffected by one another's output (cf.
Aulin, 1982, p. 94). This means by definition that the states of
the systems can be represented in the form $J=\langle J^*, J^{**} \rangle$, or, more
specifically,

$$(23) \quad \begin{aligned} J_R^*(y_R) &= s_R, \quad J_R^{**}(y_A) = x_R^{in}, \quad x_R = \langle s_R, x_R^{in} \rangle; \quad J_R = \langle J_R^*, J_R^{**} \rangle \\ J_A^*(y_A) &= s_A, \quad J_A^{**}(y_R) = x_A^{in}, \quad x_A = \langle s_A, x_A^{in} \rangle; \quad J_A = \langle J_A^*, J_A^{**} \rangle \end{aligned}$$

Thus we have divided the coupling operator J_R into two parts J_R^*
and J_R^{**}, and in the same way J_A into J_A^* and J_A^{**}. After we have
defined the states s_R and x_R^{in} the assumption of separability
means that $x_R = \langle s_R, x_R^{in} \rangle$; and similarly with respect to **A**. Separability then means possibility to separate, with respect to the
coupling operator, the effect (s) which comes from within the
system from the effect (x^{in}) which comes into the system from
the outside.

On the basis of assumption (**23**) we can now define the total
growth functions G_R, G_A, G_{in}, and G_{out}:

$$(24) \quad \begin{aligned} s_R' &= J_R^*(V_R(x_R)) = G_R(s_R, x_R^{in}) \\ s_A' &= J_A^*(V_A(x_A)) = G_A(s_A, x_A^{in}) \end{aligned}$$

and

(25) $x_R^{in'} = J_R^{**}(V_A(x_A)) = G_{in}(s_A, x_A^{in})$
        $x_A^{in'} = J_A^{**}(V_R(x_R)) = G_{out}(s_R, x_R^{in})$.

Here G_{in} stands for the impact of the social collective A on the world R, whereas G_{out} stands for the reverse impact of the world on the collective. (We shall assume that in our applications G_{in} and G_{out} really do represent causal connections, although we need not analyze the matter in more detail here.)

We can now give the following schematic illustration of the situation. In it the process starts with the inputs s_R, s_A, x_R^{in}, and x_A^{in} yielding the primed values s_R', s_A', $x_R^{in'}$, and $x_A^{in'}$ of (24) and (25) as its first-round outputs. The process then continues by having the first-round outputs as its second-round inputs, and so on. The following figure shows explicitly the first-round situation:

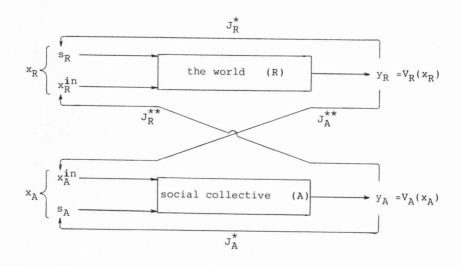

Diagram 8.1.

This figure gives a kind of macro-description of scientific change. A micro-description can be obtained once we manage to specify the elements of this figure, especially J_R^{**} and J_A^{**}, but also the feedback functions J_A^* and J_R^*. J_R^{**} can of course be connected with the performance of experiments and J_A^{**} with the making of observations (cf. causal theories of theory formation); J_A^* is connected with A's learning on the basis of experience. Needless to say, a precise description of these elements is a matter of extreme complexity. What we shall do in Chapter 9 is, in any case, to attempt to make more precise some aspects covered by the above general diagram.

CHAPTER 9

THE GROWTH OF SCIENTIFIC KNOWLEDGE

I. Truth and Explanation in the Context of Scientific Growth

1. We have claimed earlier (especially in Chapter 6) that the
notion of truth is ultimately an epistemic one. Truth inescapably depends on one's viewpoint and background knowledge (or
something like that). In what follows we shall take this for
granted and go on to investigate the relation between truth and
(best) explanation. As against some other philosophers I shall
not try to argue that (epistemic) truth and best explanation are
identical notions. But I do claim that it is reasonable to think
that true theories and best-explaining theories coincide.

As I will be dealing with very broad topics it is not always
possible to go into details. Furthermore, I will have to rely on
some of my earlier publications on these matters. I will, however, make an effort to summarize briefly some of my previously
defined key notions so as to make the present chapter relatively
self-contained.

2. In what follows I shall examine on a general level the growth
of knowledge in terms of increasingly better explanations,
having in mind the scientific realist's point of view in particular. One central assumption in this point of view is the
assumption that the sciences aim at giving ever more exact and
truthlike pictures of the world. We must keep in mind, though,
that according to (internal) realism, there is no concept-free,
viz. non-conceptual but yet cognitive knowledge (and truth) of
the world. Thus all inquiries into truth and explainability are
to be tied to one conceptual scheme or another. As the notion of
conceptual scheme is at bottom epistemic, also truth is an
epistemic notion.

We shall claim below that a best-explaining theory is a
true theory (where both are understood to be relative to some
conceptual scheme). At the same time it is epistemically ideal

relative to that conceptual scheme (cf. Chapter 6, Section I). Moreover, we can take these notions, when suitably explicated, to be equivalent with Sellars' notion of true$_a$, at least asymptotically. Thus we can claim that the following statements about a scientific theory T are, relative to its conceptual scheme or language, equivalent (though not perhaps for conceptual reasons only):

(1)   (a) T is (factually) true;
      (b) T is epistemically true (or ideal);
      (c) T is the (or a) best-explaining theory (at least asymptotically);
      (d) T is true$_a$.

Of the theses in (1), (a) represents an unanalyzed, presystematic notion of truth which is at least to some degree a correspondence notion (cf. the classical saying **"Veritas est adequatio rei et intellectus"**). The factual adequacy of a theory is understood to be some kind of adequate correspondence between the theory and the portion of the world represented by it. As I have argued in Chapter 6, to a large extent following Sellars (1963a), (1968), such adequacy is adequate picturing, and a theory cannot (at least ideally and asymptotically) be factually true unless it pictures that portion of the world correctly. On the other hand I claimed, and showed, in Chapter 6 that the concept of truth in the general sense (a) is epistemic in that it necessarily requires background knowledge. The equivalence of theses (a) and (b) is in its essentials based on this fact. We can say that (a) and (b) can be explicated in such a manner that they turn out to be equivalent, partly on conceptual grounds, partly on the basis of general rationality assumptions (cf. (17) and (22) below). The same can be said of the equivalence between (d) and (b). The best concise way to put it is perhaps this. In Chapter 6 we attempted in fact to show that (d) explicates (a); cf. the thesis (**E**) defended in that chapter. To the extent that is so the equivalence of (d) and (b) can be derived from this result and the equivalence of (a) and (b).

Let me lay special emphasis on the fact that the equivalence of (c) with the other claims above can only hold if T is a scientific theory (here: a conjunction of general lawlike sentences in a suitable pragmatically adequate formal language or, alternatively, set-theoretical language - cf. the state-space

account of theories). For instance, Rosenberg (1980) has argued
that best explanation in this case is precisely truth in the
sense of correct assertability, viz. in the sense (d). In Sellars' system, on the other hand, the equivalence of (c) and (d)
can be justified on the basis of the explanatory role of the
material rules of inference (see Sellars, 1968). I shall not
here examine in closer detail these approaches. But I shall
later argue that (c) can indeed be closely related to (a), (b)
and (d), as claimed. Note here that asymptoticity in claim (c)
refers to a situation in which the world has in a sense been
investigated exhaustively within the bounds of the conceptual
scheme employed.

Let us next have a closer look at the growth of knowledge
as processed by scientific theories. Let us consider a community, say A, of researchers who act together with the general
goal of finding out what the world is like. Let the action, say
U, performed by the community A be that of replacing a predecessor theory T_i by a successor theory T_{i+1}. How can such a
social, multi-agent action to be justified? Assume that these
agents first act within the bounds of the theory T_i (e.g. Newtonian mechanics) in something like the normal-scientific way
described by Kuhn. The application of the theory to some phenomena, say $E_1,\ldots,E_n$, proves to be impossible, however. We shall
then say that $E_1,\ldots,E_n$ are anomalies for our theory.

The scientific community A then produces a new theory candidate T_{i+1} (e.g. Einstein's theory) which is able to explain
those anomalies and which "works" also otherwise at least as
well as T_i. We assume, in other words, that T_{i+1} explains "correctively" both the success of T_i and its anomalies. On these
grounds the rational community A forms the we-intention (group
intention) to perform U, that is, to replace T_i by T_{i+1}, and
executes the intention. (Generally, a community A can be said to
have such a we-intention if all its members have it or if either
most of its members have it or at least its leaders have it; see
Tuomela, 1984, for the notion of we-intention.) We can now think
that the members of A act in accordance with the schema ($\mathbf{PR_i}$) of
Chapter 8, where action U is precisely the replacement of T_i by
T_{i+1}. The execution of the we-intention takes place through the
performance by the members of A of their own component actions,
which together bring about U.

Next, I shall present a schema of practical inference which
deals with the transference of group intentions and especially

their post factum justification in the situation just characterized (Rosenberg, 1980, p. 178, presents a rather similar
schema). We shall assume, then, that each member of A can reason
as follows, given that we allow A to survive changes in its
members, and given that 'we' here refers to a community with
such an identity:

(2) (i)   We will have a theory which affords best explanatory
           accommodation for those R-experiences (structured in
           terms of it) with which we find ourselves.

    (ii)   We believe that our adopting theories which qualify
           as successors whenever we find ourselves with R-ex-
           periences anomalous relative to our extant (prede-
           cessor) theory facilitates our having such best ex-
           planatory accommodation.

    (iii) We will adopt a qualified successor theory whenever
           we find ourselves with predecessor-anomalous R-ex-
           periences.

    (iv)   We believe that $E_1, \ldots, E_n$ are anomalous R-experiences
           with which we find ourselves but which cannot be af-
           forded best explanatory accommodation within our ex-
           tant communally accepted theory T_i.

    (v)   We believe that T_{i+1} (and it alone) qualifies as a
           successor to the predecessor theory T_i with respect
           to $E_1, \ldots, E_n$.

    (vi)   We believe that replacing T_i by T_{i+1} is a case of
           adopting a qualified successor theory upon finding
           ourselves with predecessor-anomalous experiences.

    (vii) Therefore: We will replace T_i by T_{i+1}.

Schema (2), where 'will' is used conatively as expressing intentions, is a conceptually valid one, if corrective explanation is
understood in the sense of analysis (15) to be given later.[1]
    The broader set-up into which we think (2) can be embedded
is the cybernetic system-environment interaction situation we
discussed in subsection 8.III.7. In that situation the collective A is naturally supposed to be developing and carrying out a
research programme including all the theories T_i which (2)
speaks about.
    Let us next examine the premisses of (2) more closely. As
far as premiss (i) is concerned there are reasons to think that
the experiences concerning the domain R have been conceptualized

within the bounds of the conceptual scheme of the best-explaining theory in question. Best explanation must be understood to
be relative to what is asked and it involves, among other
things, explanation which is free from anomalies. This notion
will be clarified in question-theoretic terms in Section III
below. The notion of contribution in premiss (ii) can be understood in part causally. Despite this (ii) is close to being a
conceptual truth, because the idea of contributing to the achievement of best explanation involves the elimination of anomalies. Note, however, that each theory T_i, so to speak, concep-
tualizes its own anomalies.

        We may note in general terms here that (i) and (iii) to-
gether state community A's general strategy of action in the
discussed kind of situation. While there is no need here to
provide a more detailed commentary of the premisses of (2) here,
let me yet note that premiss (ii), as understood here, involves
the notion of corrective explanation. This notion is important
and requires further elaboration, to be given later. We shall
before that discuss other topics, beginning with the general
notion of scientific explanation.

II. A Pragmatic Account of Scientific Explanation

        I shall begin with the general idea of explanation and
accept as my starting point the question-theoretic analysis I
have presented elsewhere (Tuomela, 1980b; cf. also Tuomela,
1984, Chapter 10). According to this analysis scientific explanation can be looked upon as communicative social action which
consists of querying and answering. To obtain a brief overview
of this approach we shall examine the following tentative definition in which C refers to a particular explanatory context, P
to relevant "paradigmatic" background assumptions (cf. Kuhn),
and q to the oratio obliqua form of a question (generally, a
why-question):

(3)   Agent A **explains** q **scientifically** to agent B by producing
         a linguistic token utterance u in the context C, relative
         to the background assumptions P, if and only if
         (a) A believes that u is a scientific explanatory answer
             to the question q (in the context C, relative to P)
             or at least that u represents such an answer; and

(b) A produces u with the intention that producing it
    brings about in an intended way that B understands q
    (in the context C, relative to P) and that B's under-
    standing q comes about, in an intended way, through
    B's belief that u is a scientific explanatory answer
    to q or at least represents such an answer (in the
    context C, relative to P).

This analysis relies on the notion of a scientific explanatory
answer (se-answer, for short). We need to clarify it. What I
shall call a complete explanatory answer is a (broadly) linguistic token utterance which in a specific sense satisfies the
(complete) presupposition of the explanation-seeking question q
(i.e. the items presupposed by the question) and which is understandable to B and which also constitutes a nomological scientific explanatory argument for q. It may be mentioned here that
there are explanatory contexts requiring several answer tokens
(cf. u above) to be produced and which thus transcend our (3).
In such contexts the relevant intentions of the explainer (cf.
(b)) are broad conditional ones and amount to explanatory strat-
egies.

Let me give a simple illustration of this idea. Suppose
somebody asks the question "Why did this object a expand?". We
answer this question in our account by implicitly answering the
following so-called complete question corresponding to the above
question: (?f)(a expanded for the reason f). Let us denote this
complete question underlying our original question by q. What
you find within the parentheses is called the complete presupposition of q. An answer to q will now be given in terms of the
following argument: i) a is a piece of copper, ii) a was heated,
and iii) all pieces of copper expand when heated; therefore iv)
a expanded. Now the statement "iv) for the reason (that) i)&ii)
&iii)" counts as a (potential) complete scientific explanatory
answer to q (apart from the obvious empirical simplifications
and idealizations involved). Denoting the conjunction i)&ii)&
iii) by f_0 we can say that an se-answer serves to give a value
(such as f_0) to the question quantifier (?f); and we say that f_0
constitutes an E-argument **for** q.

Let me now reproduce my definition of a complete scientific
explanatory answer as follows:

(**4**)   A   linguistic token-statement u is a **complete scientific
        explanatory answer to q in situation C,   given P,   if and
        only if
        (a) u   is obtained from a complete presupposition (rela-
             tive   to C and P),   say (Ef)s(f),   corresponding to q
             (formalized by (?f)s(f) in its **oratio recta** form)   by
             dropping the existential quantifier and by substitut-
             ing a constant, say f_0, for f in it;
        (b) u is P-understandable in the context C; and
        (c) u constitutes a nomological argument, viz. an E-argu-
             ment,   for   q   such that this argument has f_0 as   its
             conjunction of premises.

        This definition embeds a somewhat Hempelian kind of idea of
nomological   explanation as a logical argument into a   pragmatic
context of scientific inquiry.   There is no need here to clarify
in   more   detail the technical clause a) of (**4**),   for it has   no
direct   bearance   on our present topic.   But clauses b)   and   c)
clearly are relevant.   For when we want to compare the   goodness
of   several   explanatory   answers to one and the   same   question
there   are two factors in (**4**) relevant for that,   and   they   are
the following.
        First and foremostly,   we have the notion of   P-understand-
ability in b).   In comparative terms,   the more understanding an
se-answer produces the better it is. Thus, ceteris paribus,   explanatory goodness covaries directly with the amount and kind of
understanding it produces. Secondly, we have the factor of goodness of argument related to c). The notion of E-argument in this
clause   means a nomological argument.   In the deductive case   we
are   dealing with ε-arguments and in the inductive case with ρ-
arguments. (For lack of space I can here do no better than refer
the   reader to my previous publications,   e.g.,   Tuomela,   1973,
1976,   1980b,   1981,   for   a characterization and discussion   of
these notions.) We can say that,   ceteris paribus and with   some
qualifications, explanatory goodness covaries with the degree of
nomic   expectability the E-argument in question   guarantees.   In
this   sense   deductive   explanations are better   than   inductive
ones, ceteris paribus. (However, the matter of nomic expectability is tricky - cf. low-probability explanations.)
        What   I   call a scientific explanatory   answer **simpliciter**
consists   of mentioning only a part of the E-argument in a   complete scientific explanatory answer (see Tuomela, 1980b, defini-

tion (9) for the exact notion). Thus, in our earlier example,
for instance, "Because a was heated" would be a (potential)
scientific explanatory answer simpliciter.

The notion of corrective explanation that is implicit in
our schema (2) can be partly analyzed by means of the above
notions, but it also involves other notions, to be clarified
below. Therefore we shall leave our remarks on corrective explanation till the end of the next section.

III. What Is Best Explanation?

1. Let us now proceed to the situation our schema (2), concerned
with theory-change, deals with. There we have a predecessor
theory T in L (corresponding to T_i in (2)) and its successor
theory T' in L' (corresponding to T_{i+1} in (2)). In the general
case we have perhaps only a translation function, say tr, connecting the two languages extensionally.[2] The function tr is
based on the possibility of comparing the models of T and T' -
even in cases where these models contain entities of quite different kinds; see Tuomela, 1984, Chapter 14, and note 2 above.

Let us now assume that we can specify our problem area - a
set of questions - within each language. The number of problems
or questions requiring answer in principle may be less in L than
in L', for typically the successor language L' carves up the
world in a finer and more informative way than L (cf. the discussion of monadic conceptual enrichment in Niiniluoto and Tuomela, 1973, and below in Section IV). In any case, we translate
L into L' by means of tr in the general case and thus assume
this much comparability. As defined in Tuomela (1984), tr is
truth-preserving (viz. if a sentence in L is true then its
translation in L' will also be true) and therefore our comparability assumption is nontrivial. We can also say that this
function tr expresses a principle of information retention.

Now we continue by trying to compare the explanatory power
of T' with that of tr(T) rather than T itself. Both T' and tr(T)
are in L'. (Below, we will normally call the predecessor theory
T also when it is couched in the language of the successor
theory, just to avoid the clumsy expression tr(T).) Next we
assume that we have available a common problem area for our two
theories to be compared, and by a problem (anomaly or other
problem) we here simply mean a question stated within L'.

Let me first define a notion that we will have some use for below. Let thus u and u' be se-answers to q. We shall assume from here on that the language L' can be understood in a very broad sense so that it incorporates paradigm- and context-relativity, viz. relativity to P and C; thus relativizing our analysandum to L' will conveniently take care of these elements. (We may treat or view L' as a set of inference rules, for instance, and suitably incorporate the information in P and C into them.) We then propose:

(5)   u' is **at least as good as** u **in answering** q relative to L'
        if
        (1) u' makes q at least as much P-understandable (in C)
            as u does; and
        (2) u' constitutes an E-argument for q which is at least
            as good as that constituted by u.

Note that (5) only states a sufficient condition. This is be-cause some compensatory interaction between 1) and 2) might affect necessity and thus neither 1) nor 2) by itself qualifies as necessary for the analysandum. For u' might be an E-argument worse than u but render q better P-understandable in a way which makes u and u' equally good as a whole.

Let us now say that a theory T generates a scientific ex-planatory answer u if T occurs as a premise in the E-argument that u constitutes. We then go on to define, relative to a set of explanation-seeking questions Q (representing anomalies and other problems), explanatory betterness in the case of two theories T and T' in the same language. Relating the present situation to the context of scientific growth, we may take T' to be the successor theory of a predecessor theory, whose transla-tion into L' is just T. We then propose:

(6)   T' is a **better explanatory theory** than T relative to the
        set Q of explanation-seeking questions and to L' if and
        only if
        (a) for every q in Q, if T generates an se-answer u for q
            then also T' generates an se-answer u' such that u'
            is at least as good in answering q as u is;
        (b) for some q in Q, the se-answer u' generated by T' for
            q is better than the se-answer u (if there is one)
            generated by T for q.

In **(6)** we have the notion of an se-answer being better than another one. How can it be elucidated? Obviously **(5)** is of some help here, for it displays the factors that are relevant here. Consider thus the following rather obvious suggestion:

(7)   u' is a **better scientific explanatory answer** to q than u
       relative to L' if
       (a) u' makes q to a greater degree P-understandable than
           u does and u' constitutes an at least as good E-
           argument for q as u does; or
       (b) u' constitutes a better E-argument for q than u does
           and u' makes q P-understandable at least to the same
           degree as u does.

Note that the analysans cannot quite be regarded as necessary for the analysandum in **(7)**. For suppose u' makes q P-understandable to a very high degree but constitutes a little worse E-argument than u. Yet u' can (perhaps) be regarded as better than **u**. (See **(7*)** below for a remedy.)

Note that the notion of P-understandability is a fairly global notion. It relates to a range of conceptual abilities (possessed by a person said to P-understand q) relating the question q to other questions (and answers). On the other hand, the goodness of an E-argument is a more local notion concerned with something directly relevant to q, viz. with direct nomic reasons for q.

Going back to **(6)**, note that it allows for anomalies in the case of both theories, viz. it allows for unsolved problems and unanswered questions. However, **T'** is required to solve all the problems that T is able to solve. In other words, we have built explanation retention in this sense into our system. But this may seem a strong requirement, which in any case would need an appropriate defense. For it should be noted that betterness could be achievable also by means of some compensatory mechanism. **T'** might be overall better than T even if it fails in the case of some problems that T succeeds in solving, if it still is better in the case of most problems. In view of this we should perhaps in analogy with **(7)** drop the only if-part of **(6)** - or else **(6)** is to be taken strictly as a stipulative definition.

If one prefers not to accept **(6)** as a stipulative definition the above idea of compensation can be accounted for by giving up the explanation retention property. We shall consider

one way of doing it. Suppose thus that the explanatory value of
an se-answer can be numerically measured by a measure EV so that
we   can   take EV(u) to stand for the goodness-value of u   as   an
explanatory   answer   to u (relative to L'). Obviously   the   EV-
function will have to take into account both the goodness of the
involved   E-argument   and   the degree of   P-understandability   u
confers on q. How that could be done I shall not try to investigate in the present context (but cf. Tuomela, 1980b, 1981). In
any case, were we given the EV-value of every se-answer a theory
T generates for Q we could measure the global explanatory value
(GEV) of T relative to Q (= $q_1,\ldots,q_m$) by the weighted average
of the singular EV-values, viz. by

$$\text{GEV}(\mathbf{T}) = (1/m) \sum_{i=1}^{m} k_i \text{EV}(u_i)$$

where k_i is the "importance" weight related to the answer u_i and
the problem q_i, which u_i is an answer to. In our measure GEV the
weights k_i are supposed to take into account the (possible) interdependencies between the questions q_i. Thus if a question de-
composes into, say, two subquestions, which are also members of
Q, obviously the subquestions have only something like half of
the weight of the original question, and so on. But if there are
complicated interdependencies   between   the members of Q this
weighting procedure may not be easy to make work, and if Q is
infinite it seems to be an impossible task. (Cf. the - generally
unsolvable - problem of specifying the likelihood of a conjunction of two observations in the light of theory, or background
presuppositions, on the basis of the likelihood of each one of
the observations relative to that theory or those presuppositions.)

   Furthermore, there is the problem of the theory-dependence
of questions. While all questions have their presuppositions, of
course, we must here suppose that two theories T and T' can - at
some level of abstraction, at least - be compared with respect
to a common set of problems. (Recall that we have already assumed that T and T' have been stated in the same language.)
Note, however, that viewed from the point of view of each theory
(or the answers generated by each theory) new interdependencies
(strongly theory-laden ones) between the questions generally
arise.

What is still more, note that the answer u_i may also be in various ways interconnected, and it is unclear how well the weights k_i may be employed to handle that in the general case. Thus two answers might assign separate causes to their explanandum-events or they might assign the same common cause to the two events - and this is all relevant to the explanatory power to the theory in question. But we cannot here dwell longer on these issues, which in complicated cases seem to have at best in casu solutions. Let us hence accept that our measure GEV must be taken with at least a grain of salt, and consider the following definition:

(6*)   T' is a better explanatory theory than T relative to the set Q of explanation-seeking questions and to L' if and only if GEV(T) < GEV(T').

I will not here discuss the approach (6*) further except for remarking that it does not - for better or worse - satisfy the explanation retention property, contrary to (6). It suffices for our purposes below to accept either (6) or (6*) for our analysis of the notion of better-explaining theory.

In view of the above we may now consider replacing (7) by

(7*)   u' is a **better scientific explanatory answer** to q than u relative to L' if and only if EV(u) < EV(u').

Note that (7*) may be combined not only with (6*) but with (6) as well.

Given (6) or (6*), we can say that if a theory T' is a better explainer than a theory T then clearly T' is a better problem-solver than T - problems understood in terms of questions. We can also say that in this case T' is in a pragmatic sense more informative than T. Note, too, that our (6) and (6*) define strict betterness: if T' is better than T then T cannot be better than T'.

We can now proceed to our definition of the notion of best-explaining theory in terms of (6) (or, if you prefer, (6*)):

(8)   T' is a **best-explaining theory** relative to Q in L' if and only if for all T, T' is a better explanatory theory than T relative to the problem-set Q.

If we define explanatory betterness in terms of (6) in (8),
the best-explaining theory T' will be unique. But if we use (6*)
this theory T' need not be unique, for there may be ties. Note
that (8) idealizingly concerns all the logically possible theories in L'. Therefore it can safely be assumed that the best-
explaining theory is capable of answering all the answerable
problems in Q even when Q exhausts all the problems statable and
answerable within L'. Indeed, in many cases when discussing
best-explaining theories we might want Q to be the set of all
problems (or at least all problems with correct presuppositions)
in L'. That of course would amount to quantifying universally
over Q in (8) and would mean omitting relativization to Q.

The following definition gives a weaker notion of best explanation:

(9)   T' is a **best-explaining theory** relative to Q in L' if and
        only if for all T, T is not a better explanatory theory
        than T' relative to Q in L'.

(9), contrary to (8), allows that some other theories of L' are
as good as the best-explaining theory in the sense (5). While
(8) and (9) both accept the possibility of several best-explaining theories they thus differ in this particular respect. We
need not make a definite choice between (8) and (9) for the
purposes of this chapter, even if I prefer the more clear-cut
notion (8).

As above, in (8) and (9) the relativity to language L' is
meant to reflect relativity to the "paradigm" P (and, where
appropriate, to the context C), and thus we are working here
with an extremely wide notion of language, for convenience. Note
that if we could find a theory T' which is a best-explaining
theory for every possible scientific language then we would have
a best-explaining theory in a language-independent and absolute
sense. But it seems impossible to make good sense of the quantifier 'for every language' - the class of all languages is
surely an open class due to the ever-present possibility of
conceptual innovation.

Another issue about our definitions worth mentioning is
that our notions of better explanation and best explanation are
non-epistemic and non-doxastic. We could, alternatively or in
addition, have formulated the above definitions in terms of what
the explainer believes or has good reasons to believe about the

goodness of the se-answers he is giving. This kind of definitions would be doxastic and epistemic, respectively; but it has
not been necessary to formulate such counterpart definitions
explicitly above.

The above account of better and best explanation can also
be applied to corrective explanation - a notion to be defined by
(15) below. Then we must in effect deal with approximative E-
arguments and that will take place in terms of the notion of a
theory approximating another theory.

2. Let us consider still another approach to analyzing best explanation. Thus consider the following definition of the notions
of the deductive explanatory power (DEP), inductive explanatory
power (IEP), and total explanatory power (EP) of a theory T:

(10)   (a) DEP(T) = $\{S/\varepsilon(S,T)\}$
        (b) IEP(T) = $\{S/\rho(S,T)\}$
        (c) EP(T) = DEP(T) $\cup$ IEP(T)

(a) defines the class of statements S for which T together with
some auxiliary statements is capable of giving a deductive explanation in the sense of providing an ε-argument for S. Here S
should be regarded as either a fact-describing sentence or a
law-describing sentence, suitably characterized.

In clause (b) we deal with inductive rather than deductive
arguments and thus use the notion of ρ-argument, viz. (minimal)
inductive argument. Clause (c) takes the union of the previously
defined two classes to characterize the total explanatory power
of T. Note that we have not explicitly relativized (10) to the
problem set Q but that can easily be done when needed (cf. below). (10) is implicitly relativized to the language L', as T
and T' both are in L'.

To comment on the connection between (6) (or (6*)) and
(10), we first note that in (10) we are dealing with explanatory
arguments, which are roughly of the kind "S because T (and the
auxiliary assumptions)". While such arguments have the form of
se-answers they are not yet se-answers, for they are not concerned with the central notion of P-understandability at all.
Thus the notions defined by (6) and clause c) of (10) cannot be
equivalent. But these notions are not unconnected, as we shall
see below in some detail. Both are connected closely with ex-

planatory unification, for example, as better explaining theories tend to be connected with larger EP-classes. (Recall here
Hempel's emphasis on unification, see e.g. Hempel, 1965, pp. 343
- 345, and 1966, p. 83.)

Consider now the following suggestion, claimed to be true
in view of what the goodness of E-arguments in (6) involves:

(11)   If T' is a better explanatory theory than T (relative to
        Q and L'), then EP(T) ⊂ EP(T').

I would claim that (11) is at least ideally the case, for T'
cannot be better than T (in the sense of (6)) unless it explains
all that T does and a little more. (Cf. formula (7.6) of Tuomela, 1973, for a more detailed explication of this idea.) The
converse of (11) obviously does not hold true without strong
qualifications, as the class EP has been defined without reference to the notion of P-understandability. But if we could
assume T' and T to be equal in the degree of P-understandability
they yield, the converse of (11) would also be true.

Another interesting property that we can require a better
explaining theory to have in view of (11) and what I have elsewhere said of E-arguments is that T' should not be to a greater
degree an ad hoc -theory than T is (cf. Hintikka, 1973b, Tuomela, 1973, and above Chapter 5). By adhockery I here mean especially so-called theoretical adhockery (cf. formula (7) of
Chapter 5). The denial of such adhockery entails that when a
(mature) theory is extended and applied to explain phenomena in
a particular domain, its theoretical core content should not increase in that expansion process (viz. the theory is not allowed
to be substantially changed merely in order to fit a new domain). Let me state this property or criterion of adequacy for
better-explaining theories somewhat vaguely as follows:

(12)   If T' is a better explanatory theory than T (relative to
        Q and L'), then T' exhibits less theoretical adhockery
        than T does when applied to a new domain.

While the above two desiderata for good explanatory theories relate primarily to the properties of E-arguments, it
should be obvious that the property (11) is directly connected
to the unifying power a theory has and via this to the degree of
understandability it yields concerning the problem the question

q expresses. But also (12) concerns the unifying power of a
theory: The more ad hoc a theory is, the less it obviously can
unify in a proper sense.

3. Let us next discuss scientific understanding briefly. Understanding is an ability. If a person understands something, he is
able to make relevant inferences concerning the matter, discuss
it intelligently (ideally at least!) and informatively, and (in
many cases) to manipulate the world in a relevant way. In Tuomela (1980b) I have spoken about the reason pattern (or causal
pattern) and the hermeneutic pattern of understanding; these are
understood to cover jointly the mentioned aspects of understanding. Without going deeper into that here let me just emphasize that scientific understanding involves fitting the explanandum rationally into a broader nomological network of laws,
theories, and facts, which often involve "deeper" underlying
theoretical scientific entities (cf. Tuomela, 1973, Chapter VII,
on this). Thus some degree of (deeper) unification as well as
systemicity are always involved in scientific understanding.

Scientific understanding is a very complicated notion.
While I have above taken up some central features involved in
it, a great many others could have been mentioned and discussed.
Thus, we might consider, for example, such proposed criteria of
adequacy for explanations as consilience, simplicity, stability,
originality, heuristic power, and a host of pragmatic and
strongly context-dependent factors and try to relate all these
to our present discussion. We shall not, however, do it here but
concentrate on an aspect that has not yet been properly emphasized, viz. truth.

To do that, let me start by restating the partial technical
characterization of paradigm- and context-dependent scientific
understanding given in Tuomela (1980b):

(13)   A person B (scientifically) understands q in the situ-
        ation C, given P only if
        (a) q is a sound question and B believes it is;
        (b) B has at least some acquaintance with and some con-
            ceptual mastery of the items mentioned in, or presup-
            posed by, q;
        (c) B knows an answer u to q that is correct in C, given
            P;

(d) B has some acquaintance with and some conceptual mas-
    tery of the items mentioned in, or presupposed by,
    the answer u (or, if u is not a complete se-answer,
    in the complete se-answer u* which corresponds to u).

The above analysis of understanding is related to actual
rather than potential understanding and this feature is reflected in the use of the phrase 'correct' in c). When we are
discussing the best explanation of the world in the fullest
sense it is obviously actual (or true) explanation and actual
understanding in something like the mentioned sense we must be
concerned with.

The best-explaining theory T' can be seen to be involved
with actual scientific understanding as follows. Suppose (13) is
satisfied in the case of some q. Then the se-answer u must be
correct rather than, for instance, merely believed to be correct. Thus if a child asks "From where did our new baby arrive?"
the answer "A stork brought her." might satisfy it, but this
answer does not make the child understand from where the baby
arrived in the sense of actual (scientific) understanding. This
is because the answer is not correct. Analogous examples would
be the attempts to explain combustion phenomena by reference to
phlogiston or various happenings by reference to God's will or
Zeus' thunderbolts. For even if such examples could be elabor-
ated to become some kind of potential explanations (which can be
doubted), they would fail as actual explanations and hence as
best explanations simply because there are excellent reasons to
think that there are no such things as phlogiston, Zeus or God.
So correctness must be required of best explanations.

What is it for an se-answer to be correct? Obviously it
must involve that in the corresponding complete se-answer corresponding to that se-answer the premises of the involved E-
argument must be correct. But an E-argument is nomological, viz.
it involves a law or theory. Now surely the best scientific
understanding of the world must involve that the theory in
question here is the best-explaining theory T'.

Thus we have arrived at the requirement that the best-
explaining theory T' must be correct. Correctness again can be
equated with truth in some presystematic sense. But as I have
argued elsewhere (especially in Chapter 6 above), this presystematic sense makes truth epistemic and viewpoint-dependent. The
next step in the argument is to claim that the best-explaining

theory T' is not only epistemically true but true in a suitable correspondence sense (viz. picturing) as well.

We shall below in Section IV work with theories formalizable within first-order predicate logic. Each such theory, which is a conjunction of axioms (general sentences), can be transformed into its distributive normal form (in the sense of Hintikka). As a consequence each true theory, say T^*, can be expressed as a disjunction of constituents, C_i, which are maximally informative generalizations. Thus T^* is logically equivalent with a disjunction of such constituents such that the true constituent, say C^*, will be one of them. Within our present framework it is then true that epistemic truth (in the sense of (17) and (19) below) involves the true constituent C^* - and C^* will be argued to be true in a correspondence sense as well. It follows that the best-explaining theory T' is and must be correct in a sense involving correspondence truth.

Another way of showing that correctness involves picturing or correspondence truth would go via Sellars' notion of (correct) semantic assertability (cf. Chapter 6). For the correctness involved in (13) would then be argued to be just correct semantic assertability in the Sellarsian sense. Given that, we can continue with the claim that the latter notion entails picturing truth in the case of singular descriptive statements as well as the corresponding constituents (see below (22) and Chapter 6 for details).

We have above sketched two arguments to the effect that correct explanations involve picturing truth and that best explanations must involve correctness. Thus we have given reasons to think that in the claim

(14) T' is the best-explaining theory if and only if T' is true (in the sense involving maximal informativeness)

the implication from left to right is true. How about the other implication? Could T' be true without being the best-explaining theory? Suppose T' is a logical tautology. Then it would be true (in a logical sense) but surely it would not explain anything. But our (14) excludes logical tautologies from the true sentences we are concerned with. (14) requires full factual truth about the domain in question, for that is what the maximal informativeness amounts to.

But given this, we can regard also the implication from

right to left in **(14)** as true. Why? Let us consider the matter
within our above framework, for simplicity. We are then dealing
only with constituents, for only they are maximally informative
in the required sense. But as only one of the constituents can
be true (in a given language), T' in **(14)** must be regarded as
just the true constituent C^* (or a sentence logically equivalent
with it).

We now have to argue that C^* must be just the best-explaining theory. How do we do that? Why on earth should we think that
the maximally informative true theory is the best-explaining
theory? The theory C^*, due to its truth correctly reflects all
the causal connections and invariances in the world, relative to
its conceptual framework (cf. **(22)** below). But, we may ask, what
else can the best explanation possibly do but that same thing as
well? Thus, if the best E-argument serves to make the explanandum event (or regularity, as the case may be) nomically
expectable that must in part be in virtue of its being true, we
may argue.

However, the matter is not so simple. We recall that best
explanation involves not only an E-argument but P-understandability as well. Could a maximally informative and true theory
in a language incorporating the information that P and C represent fail to be maximally P-understandable? It seems to me that
if we are dealing with even moderately rational scientists in
our P-community we should answer this question negatively. For
rationality in this scientific context can plausibly be taken to
involve the use of reliable and self-corrective means for getting at correct descriptions of the world and accordingly at
correct se-answers (to Q-problems). It thus seems that at least
in the long run rational scientists possessing the true theory
are bound to achieve the best achievable understanding of the
world (recall **(13)** and our remarks related to it). So, without
here pursuing the matter further, I will assume - at least
tentatively - that the implication from right to left in **(14)**
holds true.

Let me emphasize here again that our notions of best-explaining theory and of (maximally informative) truth (epistemic
truth being conceptually the key notion) are relative to the
paradigm P and the context C, the former explicitly and the
latter implicitly. But both notions are to an equal degree view-
point- dependent in the mentioned sense. As in effect noted,
there seems to be no way to get completely rid of this dependence, nor should there be.

4. Let us now return to the schema (2) and to the notion of corrective explanation in it. It has to do, first, with the fact that the theory T_{i+1} (corresponding to the better-explaining theory T' above) must be able to give acceptable answers to all those why-questions which are connected with the relevant T_i-anomalies $E_1, \ldots, E_n$ and which are couched in the language and conceptual scheme of T_{i+1}. Secondly, T_{i+1} must also be able to explain the success of T_i and hence to answer satisfactorily all the why-questions answered acceptably by T_i. Recall nevertheless that both theories formulate the questions on the basis of their own conceptual schemes. To put it somewhat crudely, theory T_{i+1} will on the basis of this be essentially superior to T_i as to its explanatory power over the phenomena dealt with by T_i and the anomalies $E_1, \ldots, E_n$. (This claim holds true at least if T_{i+1} does not create numerous - and serious - new anomalies relative to those of T_i.)

I have given an account of corrective explanation elsewhere (see Tuomela, 1984, Chapter 14). Here I confine myself to sketch its basic features. For simplicity, I shall leave out reference to the anomalies $E_1, \ldots, E_n$ (assumed to be included in the class Q of relevant problems) and assume that T_{i+1} manages to explain all the T_i-anomalies. Condition (a)(i) in the following analysis is to be understood to incorporate these simplifying assumptions. As before, we shall assume that the explanation takes place in or relative to the language or conceptual framework of the successor theory T_{i+1}. This relativity is left implicit in the following definition:

(15)   Theory T_{i+1} **explains correctively** theory T_i if and only if
        (a) there is a translation function **tr** from the conceptual scheme (or language) of T_i into the conceptual scheme (or language) of T_{i+1}, and an auxiliary hypothesis H and a theory T_i^* (both in the language and conceptual scheme of T_{i+1}) such that
            (i) T_{i+1} jointly with H explains T_i^*, and
            (ii) T_i^* is an approximation of $tr(T_i)$, viz. the translation of T_i into the conceptual scheme (or language) of T_{i+1};
        (b) T_{i+1} is a better explanatory theory than T_i.

Explanation in schema (15) means giving an explanatory argument (E-argument), in the first place. An E-argument is not yet a scientific explanatory answer in the sense specified in definitions (3) and (4). But it is not so far from being one. We could indeed have formulated clause (a) in that terminology and required an answer of the form "T_i^* because T_{i+1} and H" to be an se-answer in clause (a)(i). In this way we can tie up schema (15) with our earlier account of explanation. Given this connection, we can safely proceed in terms of (15). (But, if necessary, when using (15) we may assume that the so-called pragmatic factors in schema (3) bring about no essential alterations in the situation and that we hence can confine ourselves to the analysis of explanatory arguments.) Apart from some clarifying remarks, I shall not give a more detailed analysis of the contents of schema (15). (See Tuomela, 1984, Chapter 14, and note 2 above on the difficult notion of translation.)

In clause (b) reference to Q has been omitted but otherwise it should be understood basically in the sense (6) (or (6*)). However, there is the difference that (6) literally applies only to the comparison of T_{i+1} with T_i^*. To get to (b) we must make additional moves. First, there is the move of translating T_i into the language of T_{i+1}. (Note that $tr(T_i)$ corresponds to the theory T in (6) and (6*).) Secondly, there is the transition from $tr(T_i)$ to T_i^*. The motivation for introducing T_i^* is to explicate approximative explanation: T_{i+1} approximatively explains $tr(T_i)$ if it strictly explains some theory T_i^* which $tr(T_i)$ is an approximation of.[3]

What the mentioned two moves involve in terms of explanatory power in general is a difficult problem we cannot try to solve in this chapter. (Note that while clause (a)(i) may be taken to entail that T_{i+1} is a better explanatory theory than T_i^* it does not without further argument entail clause (b) on the basis of our earlier definitions.) Still it seems reasonable to assume, and to adopt as a kind of criterion of adequacy, that, given a truth-preserving notion tr of translation and a relatively strict notion of approximation, it holds that if T_{i+1} is a better explanatory theory than T_i^* (relative to Q and hence $E_1, \ldots, E_n$) then it is also better than $tr(T_i)$ and, given the adequacy of tr, also in an indirect sense better than T_i.

Note, furthermore, T_{i+1} may explain T_i^* only inductively.

This together with the mentioned leaps from T_i to $tr(T_i)$ and from the latter to T_i^* leaves room for several kinds of deductive inconsistencies between T_{i+1} and T_i as well as between T_{i+1} and the various presuppositions of not only the conceptual framework of T_i but that of T_{i+1}, too. Thus, for instance, T_{i+1} may be able to correctively answer a question with a grossly false presupposition by answering a question with a more truthlike presupposition such that the former presupposition anyhow approximates the latter; cf. the relation between T_i^* and $tr(T_i)$. But we shall not here make an attempt to systematically botanize such inconsistencies and discrepancies.

Recall that our basic schema (2) involves corrective explanation. We can now say that if the soundness of (2) and (15) is accepted then science progresses to ever better-explaining theories about domains of inquiry. However, not even this condition gives a priori guarantee to the claim that science ultimately ends up with the best-explaining theory (i.e., there is no necessity that there be such a theory), and this is what is reasonable to think of the matter. Note that all this tallies well with the fact that according to the growth function $T_A^{t\tau}$ science actually can grow indefinitely without there being any limit science with its best-explaining theories.

IV. Inductive Logic, Epistemic Truth, and Best Explanation

1. Given our account of best explanation and of comparative explanatory power we shall now proceed to a somewhat technical discussion of epistemic truth and best explanation within the context of formal first-order languages in an inductive context. We shall examine a slightly simplified case in which several relevant factors are assumed to stay constant. Let us consider a situation of growth of knowledge in which the language of the theory T_i is $L(\lambda)$, and where λ is a set of scientific predicates (here monadic).[4] Thus T_i only contains λ-predicates. The language $L(\lambda)$ can in principle take into account adopted meaning postulates and constitutive principles by ruling out the sentences (and possibilities of classification) they deny. We then assume that the language of the theory T_{i+1} is $L(\lambda \cup \{R\})$, i.e. a language which has been obtained by adding to the language $L(\lambda)$ a new monadic predicate R. We have here to do with growth of knowledge based on simple enrichment of a conceptual scheme.

(Note that there is another simplifying feature here: the ontologies of the two theories are essentially the same; the case of
theories with differing ontologies is examined, e.g., in Tuomela, 1984, Chapter 14.)[5]

The theory T_i can thus be redescribed in the language
$L(\lambda \cup \{R\})$, and hence be compared directly with T_{i+1}. In other
words the redescribed theory T_i is in fact the theory $tr(T_i)$ of
(15). We thus ground the comparison of the two theories T_{i+1} and
T_i on the comparison of the theories T_{i+1} and $tr(T_i)$. (For simplicity, we shall below call the predecessor theory T_i rather
than $tr(T_i)$.) We can then apply to this case the system of inductive logic proposed by Hintikka, in the form it has been
further developed in Niiniluoto and Tuomela (1973).

Each general sentence S of the language $L(\lambda \cup \{R\})$ can now be
represented as a disjunction of the constituents C_i of this
language: $\vdash S \equiv C_1 \vee ... \vee C_n$, for some selection of constituents
$C_1,...,C_n$. If the number of λ-predicates is m-1, the number of
the constituents of the richer language is 2^K, with $K=2^m$. We can
illustrate the situation by saying that our language partitions
the world into K "cells" or types of individuals. Each constituent is a strongest (maximally informative) possible description
of such a world, and the description says that in the world a
certain number w of the cells or types are exemplified, while
the rest K-w are unexemplified.

Next, we shall assume that, relative to the research community A, we have defined a distance measure d_A for the sentences
of the language $L(\lambda \cup \{R\})$ such that $0 \leq d_A \leq 1$. (See e.g. the possibilities presented in Niiniluoto, 1977, 1978, and Tuomela,
1978b.)[6] In this way we can further define a kind of coherence
measure for sentences. Thus, omitting reference to A, we have:

(16) $k(S,S') = 1-d(S,S')$

represents the degree of **epistemic coherence** or similarity between the sentences S and S' by help of the distance measure
$d(S,S')$. This coherence is in part based on purely formal, syntactic similarity, in part on epistemically interpreted free
parameters (see Tuomela, 1978b, 1980a; I shall discuss the epistemic interpretation of the measure d later). Sentences S and S'
thus both represent potential or partial knowledge (from the
point of view of the community A). Once we have defined episte-

mic probabilities for these sentences we can talk of degrees of knowledge.

Pursuing this line, we shall assume that we have defined, for the sentences of the language under scrutiny, and in accordance with Hintikka's inductive logic, an epistemic probability measure P_A where the index A - which will often be suppressed below - refers to relativization to the community A. P_A represents degrees of knowledge. It satisfies the standard probability axioms as well as some symmetry and regularity principles (see e.g. Niiniluoto and Tuomela, 1973, and Hintikka and Niiniluoto, 1976).[7] We shall pay particular heed to the conditional probabilities $P(C_i/e)$ which tell us the epistemic probability of a constituent C_i relative to some evidence e. Here e is assumed to effect an unequivocal partitioning of the observed individuals into types. As a general epistemological warrant for the measure P we can employ some suitable theory of knowledge, e.g. Sellars' causal one.

Given the above assumption, we now reason as follows. The conceptual scheme represented by the language $L(\lambda \cup \{R\})$ is at bottom an epistemic system which comprises the rules which govern human action (cf. the Sellarsian world-language, language-language, and language-world -rules), and the practical inferences which ground or give a warrant for these rules. Part of all this is structurally built into the language $L(\lambda \cup \{R\})$ itself as well as into its constitutive postulates, part is taken into account by the probability measure P and the coherence measure k.

It is not in fact altogether farfetched to think that men ideally aim at maximizing their expected epistemic utilities - this is but a generally accepted rationality principle. Why could we not, then, think that epistemically ideal truth can be and should be analyzed in this fashion? Isn't epistemic truth precisely what is, ideally, expected to be epistemically best? We shall answer this question affirmatively and give further elucidation and motivation for this in the course of our subsequent discussion.

We can then give the following definition for the epistemic truth of a sentence S in the language $L(\lambda \cup \{R\})$:

(17)   S is **epistemically true** relative to a research community A and the language $L(\lambda \cup \{R\})$ if and only if S maximizes the expected epistemic utilities of the community A.

Here we assume that the community A uses the language $L(\lambda \cup \{R\})$ (and, given this, we could have omitted explicit relativization to A in (17)). We shall make the further simple assumptions that the only epistemic utility of the community A is epistemic coherence in the sense of measure k_A, and that expected utilities are formed by help of the measure P_A. Insofar as the probability measure P_A is at all an adequate measure of the degrees of knowledge of the community A (and at the same time also, perhaps, of the degrees of its rational, epistemically warranted beliefs), its maximum value of course must stand for factual truth, in fact precisely (something like) picturing truth (cf. Sellars, 1968, and Chapter 6). To put it in the Sellarsian terminology, P_A primarily takes into account the relationships or rules between language and the world. On the other hand, the measure for coherence k_A takes into account the central epistemic feature of truth - its dependence on the whole conceptual scheme with the relevant background information (recall Chapter 6). Let us see where this assumption leads us.

We shall start with the formula for expected epistemic utilities, omitting again the subindex A:

$$(18) \quad EU(S/e_n) = \sum_{i=1}^{K} P(C_i/e_n)k(S,C_i),$$

Here the subindex n refers to the number of "observations" or individuals contained by evidence e_n. The expected epistemic utility of a sentence S is, then, calculated as follows. The epistemic coherence of sentence S, relative to the language $L(\lambda \cup \{R\})$, with each basic element or constituent C_i, is measured with the help of k. The higher $k(S,C_i)$ is, the greater is the epistemic utility of the sentence. These utilities are then weighted by the degree of knowledge (P) of each constituent. This degree is understood to be relativized to all the available evidence e_n which can be fully represented in the language $L(\lambda \cup \{R\})$.

We now explicate definition (17) as follows:

(19)   S is **epistemically true** with respect to a research community A and language $L(\lambda \cup \{R\})$ if and only if S maximizes
        formula (18), when $n \to \infty$, for all generalizations of the
        language $L(\lambda \cup \{R\})$.

In Hintikka's inductive logic there is a unique constituent, C_c, such that, when the amount of evidence grows without limit, viz. $n\to\infty$, then $P(C_c/e_n)\to 1$. This constituent C_c says that the world is of the same type as the asymptotic evidence: exactly those c kinds of individuals are exemplified in the world as in the evidence. Here c expresses the said number of the kinds of individuals. We now have that

(20) $EU(S/e_n) \to k(S,C_c)$, when $n\to\infty$,

and

(21) $EU(C_c/e_n) \to 1$, when $n\to\infty$, with a constant c.

A sentence, say S^*, which maximizes formula (18) and hence satisfies formula (19) must therefore be logically equivalent with constituent C_c relative to asymptotic evidence. Furthermore, the following equation must hold: $k(S,C_c)=1$. (We could, as far as I can see, require more generally even that $k(S,C_c)=1$ if and only if $S=C_c$.) Formulas (20) and (21) show, in a sense, that in the asymptotic case the import of the "internal" background knowledge (but not of the "external" background knowledge required for language use) of a language system for an epistemically true theory in the end vanishes - the crucial role will be allotted to its probability.

2.   Although we have, in pursuing the above line of thought, had monadic predicate logic in mind, the results (20) and (21) are valid for the whole first-order predicate logic. They show in an intriguing way the possibility of connecting epistemic truth with picturing truth asymptotically (i.e., when $n\to\infty$). For the result (21) shows that constituent C_c is epistemically true in the sense of definition (19) and that every epistemically true sentence S^* is logically equivalent with C_c, relative to asymptotic evidence. In the asymptotic case the distance of an arbitrary sentence S from epistemic truth is, according to (20), given by $k(S,C_c)$, i.e. by the degree of epistemic coherence of S with C_c.

We can now think that the truth of evidential sentences is defined as picturing truth (see Chapter 6). Thus the evidential sentence-token 'a is of kind Q_i' can be thought to be a Sellarsian verifying token. Although evidential sentences in Hintikka's system in the end are construed as existential and "statistical" (e.g. "There are p individuals of the type Q_i"), the philosophical setting is not thereby changed in any essential way. When the amount of evidence grows without limit, it is quite natural to generalize the definition of picturing truth to cover also constituents in this asymptotic case. For all that a constituent says is how the individuals of the world are partitioned to different cells, when all this information in fact exists in the singular evidence obtained through causal picturing. Although I shall not here give any technical arguments, I will assume that unlimited (i.e. $n \to \infty$) singular description of the world through picturing is a necessary and sufficient condition for the determination of the picturing truth of a constituent C_i, and that C_c is true in the picturing sense if and only if the evidence in fact (asymptotically) exemplifies the c kinds of individuals claimed by C_c to be exemplified.

As already said, we shall, on the other hand, define the epistemic truth of constituents and other descriptive sentences by means of (19). Since (19) refers to an asymptotic case, it follows that the notions of picturing truth and epistemic truth in fact coincide in the asymptotic case. To avoid unnecessary technical complications we shall assume that sentence S has been transformed into its distributive normal form. Relying in part on this we claim that $k(S,C_c)=1$ if and only if $S=C_c$. Thus we get:

(22)   S is epistemically true if and only if S is true in the picturing sense, when $n \to \infty$.

On the other hand, it stands to reason that in the case of descriptive statements epistemic truth as defined by (17) (and (19)) is equivalent with (or, if you prefer, explicates) Sellarsian assertability truth ($truth_a$). An exact demonstration of this equivalence would require us to discuss how the world-language -rules for the collection of evidence ought to be formulated. Similarly, as we have already indicated, it would require us to formulate in the form of rules the axioms which constitute the probability measure P_A and the coherence measure k_A. Thus we would end up, after some rather non-Sellarsian arguments, with the conclusion that $true_a$ and $true_p$ are equivalent also for generalizations and not only singular sentences. This of course tallies very well with our causal internal realism (CIR) (cf. Chapter 6) and with the plausible idea that truth_a and truth_p may both obtain without this being finitely knowable.

The results presented above give a warrant for the equivalence of the claims (a), (b), and (d) of hypothesis (1). All in all, we have claimed that in the general case factual truth in a sense amounts to epistemically ideal truth, i.e., that (1)(a) is to be construed as equivalent with claim (1)(b), the latter understood as explicated above. Indeed, our analysis of (1)(b) with the help of formula (19) - which need not be congenial to all supporters of epistemic truth - shows that at least for our model languages (first-order languages) epistemically ideal truth turns out to be, on the basis of some rationality and adequacy conditions (including causal ones), equivalent also with asymptotic picturing truth. As we argued in Chapter 6, truth_a is equivalent with ideal epistemic truth - in fact the former explicates the latter. The above line of argument then shows that the notion of truth_a can in its turn be considered equivalent with the truth of the right hand side of formula (19). Thus (1)(d) is equivalent with (1)(b). So we are here dealing with two equivalent though different epistemic notions of truth (cf. also the equivalence (E) in Section III of Chapter 6).

Let us also note here that Niiniluoto (1980) has used formula (18) as an estimate of truthlikeness, but that otherwise he operates with different notions of truth and truthlikeness. In fact we can also use the same formula (18) as a kind of estimate for epistemic truth (19), in the preasymptotic case. So we may come to use adequately locutions like "S seems to be epistemically true" and "S is believed to be epistemically true". While such talk can in part be justified in terms of such asymptotic features as expressed by (21) and (22) the whole matter of estimation is still problematic in a number of ways. First, the epistemic probabilities P_A of a community A can for small values of n be such that $P_A(C_k/e_n) > P_A(C_c/e_n)$, $k>c$, and $EU_A(C_k/e_n) > EU_A(C_c/e_n)$. On the other hand the probability P_A is - as some psychological inquiries into such measures have demonstrated - a highly idealized notion as compared with the actual epistemic and subjective probabilities of the community A. The same is true of the coherence measure k and its actual counterparts. We cannot therefore without argument use such idealized notions as a basis for the estimation of truthlikeness. We may, however,

note that the project of estimating truth within the present
set-up is philosophically meaningful and non-circular because of
our result (22), even if finite evidence does not suffice for
this estimation in a strict sense.

Let me note that we can measure degrees of truthlikeness in
the language $L(\lambda \cup \{R\})$ irrespective of how we solve the problem
of estimating epistemic truth. This we can do simply by using
formula (18). Thus the value

(23) $EU(S^*/e_n) - EU(S/e_n)$

measures the degree of **truthlikeness** of an arbitrary sentence S,
when S^* is epistemically true in the sense of (19). (S^* is,
however, often or usually unknown in actual research - cf. the
condition $n \to \infty$.) When we investigate a situation of growth of
knowledge and recall the terminology of our (15), we can examine
if the difference $EU(S^*/e_n) - EU(T_{i+1}/e_n)$ in fact is smaller than
the difference $EU(S^*/e_n) - EU(tr(T_i)/e_n)$ that is, if $EU(T_{i+1}/e_n) >$
$EU(tr(T_i)/e_n)$ and if T_{i+1} thus is closer to truth than $tr(T_i)$.
Note that when $n \to \infty$, we are here according to (20) essentially
making comparisons between the degrees of coherence $k(T_{i+1}, C_c)$
and $k(tr(T_i), C_c)$.

I have above sketched a means by which the degrees of
truthlikeness and truth of successive theories can be studied,
provided we can ground the comparisons in the language or conceptual scheme of the successor theory. In the general case this
presupposes a viable specification of the translation function
tr. The above discussion deals with linguistic enrichments
(transitions from the language $L(\lambda)$ to the language $L(\lambda \cup \{R\})$,
which are relatively simple from the point of view of theory
comparison. As said, for the general case I shall refer to
Tuomela (1984), Chapter 14, and to notes 2 and 3 below.

The carrying out of such a translation obviously involves
clarifying the conceptual relations between languages (cf.
(15)). On the naturalistic side, so to say, there is the corresponding transition from picturing within one linguistic system
into picturing within another, successor system. As was clarified in Chapter 6, picturing is in a sense tied to the **ought-
to-be** -rules of a language (in the Sellarsian sense), although
it is describable as non-intentional but still goal-directed
action, independent of the rules of a (or any) language. This
opens a way to characterize, in principle at least, the degrees

of adequacy of picturing in a way which is language-independent.
Thus it gives a naturalistic means of justifying the growth of
knowledge which takes place on the conceptual side, so to speak.

3. Let us now turn to the topic of best explanation. We claimed
above that the epistemically true theory in the language
$L(\lambda \cup \{R\})$ will be precisely C_c (or at least a theory which is
logically equivalent with it). But this theory may be claimed to
give also the best explanation of the sentences of the language
$L(\lambda \cup \{R\})$. This holds at least on the condition that explanatory
power is measured by a means which satisfies the following like-
lihood-condition, and selects, in this sense, the most informative (logically strongest) of the rival theories. The likeli-
hood-condition says that when we compare two theories T_1 and T_2
for their respective explanatory powers relative to the sentences g of the language $L(\lambda \cup \{R\})$ (specifically, relative to the
sentences $g(\lambda)$ which contain only λ-predicates and which represent anomalies), and when the evidence or background knowledge
comprises the sentence e, T_2 gives a better explanation of g
than T_1 if and only if

(24) $P(g/e\&T_2) > P(g/e\&T_1)$,

where $P(g/e\&T_i)$ is the likelihood of g relative to evidence e
and theory T_i, i=1,2. Here T_2 can be a successor theory which
contains $\lambda \cup \{R\}$-predicates and T_1 predecessor theory which only
contains λ-predicates (or, to be more exact, T_1 is the counterpart theory, couched in the language $L(\lambda \cup \{R\})$ of the successor
theory, of the predecessor theory originally formulated in the
language $L(\lambda)$). More generally, we may measure the average explanatory power of T_2 and T_1 relative to all the observational
statements (at least generalizations) in the observational, or
rather predecessor counterpart language, $L(\lambda)$.
     Such measures of explanatory power (relative to sentences
$g(\lambda)$) have been examined e.g. in Niiniluoto and Tuomela (1973),
Chapter 7. Applied to Hintikka's inductive logic the measures of
explanatory power which satisfy the **likelihood**-condition give
precisely the result that C_c is asymptotically the best-explaining theory (in the likelihood-sense) in the language $L(\lambda \cup \{R\})$.
This of course is based on the result (21), which implies that
only the likelihood of the constituent C_c receives a non-zero

value when $n \to \infty$. This being so the explanatory power of other theories can be studied relative to the explanatory power of this constituent.

The notions of explanatory power and best explanation are, however, relatively complex matters. Our present investigations in any case indicate that, at least in the case of first-order languages and within inductive contexts, the best-explaining theory will be or contain the most informative true theory, viz. C_C. Note that our above treatment only depends on the goodness of inductive E-arguments, viz ρ-arguments. If we now recall our definition (8) of best explanation we notice that we should also consider matters related to P-understandability. We can, however, quickly see that matters related to understandability cannot change the validity of the above claim, for best explanation (in inductive contexts at least) involves truth regardless of the degree of P-understandability the best-explaining theory confers upon the explanandum-statements.

But let us now consider the converse statement, viz. that the (maximally informative) true theory must also be the best-explaining one. As we noticed in our discussion of formula (14), this is a somewhat moot point. We can of course trivially say that given that a theory T yields a maximal amount of P-understandability, then if T is true it must also be the (or a) best-explaining theory. For, as we have previously argued, the true theory C_C will guarantee maximally good (inductive) E-arguments. But we can say more, as we did before in Section III. A rational community of scientists will at least in the long run come to understand the world better if it possesses a true theory rather than a (perhaps grossly) false theory. This concerns both causal and hermeneutic understanding (in the sense of Section III and Tuomela, 1980b). In any case the above discussion and results lend support to the truth of our hypothesis (1) concerning the equivalence of best explanation and epistemic truth (recall the varieties of scientific realism, (MSR), (MOSR), (ESR), and (SSR), which rely on best explanation). This equivalence is not of a purely conceptual nature. Rather, it hinges in part on some rationality assumptions (recall note 7) as well as on our views on explanation.

If explanation and truthlikeness are thus tied up with the constituent C_C of the language $L(\lambda \cup \{R\})$, and if the connection is preserved in the transition to the successor language, and so on, science does in fact get closer to the truth - provided that we assume that scientists always act according to the schema (2) (or some similar schema) and hence succeed in bringing about ever better-explaining theories. The converse also holds: with the above proviso, science proceeds towards ever better explanatory theories if it gets closer to truth. (Recall here also our cybernetic explication of theory-world interaction in Chapter 8.)

There are also various kinds of formal results about, say, when some successor theory gives a better explanation of empirical generalizations and anomalies than its predecessor, in the sense of formula (15) (cf. Tuomela, 1973, Chapters VI and VII and Niiniluoto and Tuomela, 1973, Chapter 7).

In (2) and (15) we have chosen as the point of departure the requirement of ever-increasing explanatory power. If the above connection is accepted, we have as a consequence also ever-increasing truthlikeness. Note nevertheless that our assumptions clearly make it possible that science continues to grow forever and without limit, for it is possible that there always are better and better languages and conceptual schemes.

Let us now consider (2) and ask why, then, the scientific community A should infer and act in accordance with it. One general reply to the question is that doing so is in various ways **rational**. In other words, it is rational to aim at best and perhaps exhaustive explanations in science, although such explanations might never be found, on the ground of ever possible conceptual innovations. We may also say (in view of the connection between best explanation and truth) that it is rational to hold that the aim of science is to tell us what there is - **scientia mensura**. The means for striving for these goals of truth and best explanation, presented in formula (2), can also be considered rational, for it deals with transitions to better-explaining (and more truthlike) theories and in this sense to increased possibilities for understanding and controlling the world. (Let me here also refer to the relevant but quite different Kantian arguments which Rosenberg, 1980, brings to bear on his corresponding formula which, in part, resembles (2).)

In nuce: my thesis, founded on the above results, concerning progress in science is that science, rationally pursued, grows towards ever better-explaining theories and, therefore, towards ever more truthlike theories. As such this is not a factual claim concerning scientific progress, but rather a rational-normative claim of, roughly, the form "The scientific

community A ought to act rationally; and if it is rational it
acts in accordance with formula (2). As a result of this science
grows towards ever more truthlike theories".

We have consequently run into the distinction normative (or
normative-rational) versus factual, for applying the schema of
inference (2) to a community A presupposes that A is rational.
If A is not rational in the sense specified, its theories may in
fact develop in a quite different way. Whether A de facto is so
rational and whether science actually will develop towards increasingly better-explaining and truthlike theories is naturally
a problem which only future science can solve.

We ought to remind ourselves that the growth function $T_A^{t\tau}$
can be interpreted either as a factual (as we have done above)
or a normative-rational function. Neither interpretation need,
as such, deal with philosophical problems related to conceptual
schemes and comparisons of conceptual schemes, for the community
A can also be viewed, so to say, sociologically without commitment to or use of its conceptual scheme. Thus we can think that
this gives us a kind of empirical basis for scientific progress,
one which is free from comparisons of conceptual schemes (cf.
the corresponding claim about picturing above). If, namely, $T_A^{t\tau}$
can be construed empirically, we can say that the continuity of
the "life" of the community A is, on the level of action, tied
to this function, and that this continuity in no essential way
depends on comparisons between conceptual schemes (or on other
comparisons presupposed by formula (2)). In this sense we may
say that semantic or conceptual incommensurability can go together with "pragmatic" commensurability.

We can add here that the group identity of community A is
not affected by some amount of change of membership, or by its
structure undergoing minor changes, as long as the basic form of
the growth function $T_A^{t\tau}$ is retained (naturally, $T_A^{t\tau}$ is technically altered, if e.g. the number of members of A changes). In
point of fact, there is no compelling reason to construe the
agents of the community A as concrete agents. $A_1,\ldots,A_n$ can
equally well stand for social **positions** or **roles,** if you like.

V. Scientific Realism and the Growth of Science

1. I have so far said rather little about the interconnections
between scientific realism and the growth of scientific knowl-

edge. Yet talk of truth (especially picturing truth) and approach to truth, as well as of increase in explanatory power, is
typical of scientific realists. Though I have already presented
my own theory of the growth of scientific knowledge, intended to
tie up with causal internal realism, it is nevertheless instructive to have a more general look at the realist notion of the
growth of knowledge. In what follows we shall deal with the
objections to realism put forth by Laudan (1984).

Laudan takes realism at its core to be a normative doctrine
with the basic claim that the aim of science is to find ever
truer theories about the natural world. But he also claims that
the modern-day realist typically conjoins to this axiological
thesis a descriptive one, viz. that the history of science,
especially in recent times, can best be understood as an exemplification of realist ideals. While he acknowledges that the
normative claim of the realist is logically independent of the
descriptive claim, still, if the descriptive claim were false
(as he claims it is), serious doubts could be raised about its
normative counterpart.

Laudan (1984) then goes on to formulate the **descriptive**
claim of realism, viz. the doctrine of what he calls **convergent
epistemological (scientific) realism,** by which he in fact refers, above all, to Putnam's (1978) theory of the growth of
knowledge. (We have already in Chapters 6 and 7 shortly dealt
with and commented Putnam's views on these matters.) Laudan
defines convergent epistemological realism, which is to be conceived as a broadly speaking **empirical** theory, by means of the
following five theses:

(R1)   Scientific theories (at least in the mature sciences) are
       typically approximately true, and more recent theories
       are closer to the truth than older theories in the same
       domain.

(R2)   The observational and theoretical terms within the the-
       ories of a mature science genuinely refer (roughly, there
       are substances in the world which correspond to the
       ontologies presumed by our best theories).

(R3)   Successive theories in any mature science are such that
       they preserve the theoretical relations and the apparent

referents of the earlier theories; i.e., earlier theories are limiting cases of later theories.

(R4)   Acceptable new theories do and should explain why their predecessors were successful insofar as they were successful.

(R5)   Theses (R1) - (R4) entail that (mature) scientific theories should be successful; indeed, it is said that these theses constitute the best, if not the only, explanation for the success of science.

In so far as convergent epistemological realism is an empirical theory it can be thought that theses (R1) - (R4) can be subjected to empirical tests, by relying on the metaphilosophical thesis (R5). And Laudan in fact does claim that convergent realism accepts the following two abductive arguments. The first of these, which specifically supports theses (R1) and (R2), is given by Laudan in the following form (Laudan, 1984, pp. 107 - 108):

I   1. If scientific theories are approximately true, they will typically be empirically successful.
       2. If the central terms in scientific theories genuinely refer, those theories will generally be empirically successful.
       3. Scientific theories are empirically successful.
       4. (Probably:) Theories are approximately true and their terms genuinely refer.

Laudan's second abductive argument relates to thesis (R3) and runs as follows:

II   1. If the earlier theories in a mature science are approximately true and if the central terms of those theories genuinely refer, then later, more successful theories in the same science will preserve earlier theories as limiting cases.
       2. Scientists seek to preserve earlier theories as limiting cases, and generally succeed.
       3. (Probably:) Earlier theories in a mature science are approximately true and genuinely referential.

It may be reasonable to think that the premises of the arguments I and II give some support to their conclusions, if they
can be assumed to be true. (This of course depends on the exact
content given to such key terms as 'empirical success', 'approximate truth', and 'genuine reference'.) Yet (as is well
known from the theory of inductive inference; cf. Niiniluoto and
Tuomela, 1973) they do not suffice to make these conclusions
probable (but only to increase their probabilities). Laudan has
thus presented invalid abductive arguments, which a realist
should then not feel compelled to accept. What is more, except
for hinting at Putnam, Laudan does not really document the
realists who also hold the theses in the sense of accepting the
premises as factually true. I doubt if there are any, although
I believe that realism can be appended with the thought that the
success of science is somehow relevant for the validity of
realism (recall my discussion of best explanation).

   Laudan (1984) concentrates on showing the falsity or at
least vagueness of the premisses of these arguments, and operates here with a pragmatic notion of success, which is, on the
one hand, too strong to make the premises true but, on the other
hand, too weak to warrant ontological conclusions. Furthermore,
he relies heavily on historical examples - and one should keep
in mind both that history can be written in many ways and that
one can never convincingly prove or argue for such partly rational-normative theses as the above theses by means of purely
factual historical considerations. Nevertheless, in this way he
attempts to overthrow convergent epistemological realism. I
agree with Laudan to the extent that at least premiss 2. of
argument I is factually false. As to its premise 1. it is perhaps also historically false - without this being connected to
the notion of approximate truth or having much bearing on realism. (As typically basic scientific theories have no empirical
content, although they of course have factual content, a realist
can admit that scientist may sometimes fail to successfully
apply such theories to empirical phenomena even when these
theories are approximately true.)

   As to argument II, as factual hypotheses both of its premisses are false (see Laudan's, 1984, historical counterex-
amples). However, this fact has little to do with the correctness of realism, for a realist has neither good reasons nor any
compulsion to assume the premises to be true. Consequently I
shall not examine Laudan's specific (and in part faulty) argu-

mentation, nor therefore attempt to turn into a sociologist of
knowledge and amend - should it be possible to do so - the
premisses into acceptable ones. In all, Laudan's arguments I and
II, which he wants to attribute to realists, are both invalid
and unsound, and a realist has good reasons for rejecting them.

2. Instead, we have occasion to pass a few remarks on the accu-
racy of the theses (R1) - (R5). Thesis (R1), when suitably
qualified, may turn out to be a factually correct historical
hypothesis. Whether it is, depends, in part, on the notion of
truthlikeness employed. If, for instance, we accept our analysis
in Section IV of it and if the measure of coherence, k_A, depends
on free epistemic parameters, (R1) is most naturally (at least
at the present stage of inquiry) to be thought more as a principle of reconstruction than a full-blown factual hypothesis
with a presently determinable truth value. The same is true of
thesis (R2), for the notion of reference is amenable to several
distinct philosophical (and historical) interpretations, and in
addition the whole issue depends highly on how mature a mature
science should be taken to be. (Besides, Laudan has in his
presentation of the thesis dropped Putnam's qualification 'typically'.)

Thesis (R3) requires quite a few qualifications to be historically true. On one hand it is clear that it holds in many
cases (e.g., Newton's theory can in a sense be taken to be a
limiting case of Einstein's theory). The notion of reference
that (R3) employs is open to several interpretations, too. (For
instance, should we take Mendelian 'gene' to refer to DNA-
segments or not?) In any case, at least an internal realist can
refuse to incorporate the thesis (both as a normative and as a
descriptive one) into his program, for he does not exclude
radical conceptual changes, nor need he require that scientists
without exception are rational in this sense. (Note that not
even strict obedience to schema (2) implies action which necessarily accords with (R3).)

(R4) is an acceptable rationality principle when understood
as a requirement of explanatory betterness - in fact successor
theories must also explain the anomalies and not just the suc-
cesses of their predecessors (cf. schema (2) and our analysis of
better-explaining theories). About this thesis we can also say

that it has historically turned out to be false a great many times.

In Chapters 4 and 5 I have shortly commented on the explanation of the success of science by help of realism. A detailed examination of thesis (R5) presents a complicated problem, and I cannot go into it here. I shall merely refer to the connection between truth (and truthlikeness) and explanatory power which was discussed at the end of section III (cf. also the arguments in section II of Chapter 4). Essentially the same line of thought can, I think, be used to support the truth of the part of (R5) which says that mature theories are successful, for explanatory power must of course constitute a central ingredient in scientific success. (In fact, what was said in section III can easily be extended to cover predictive power, another main component in that success.) However, I do not think it is correct to say that (R1) - (R4) imply that scientific theories are (specifically) **empirically** successful, although success in a more general sense is implied. While a theory in the long run can be empirically successful only if it is true or approximately true, clearly successful empirical applications of a true theory may not be immediately available.

My general conclusion on the theses (R1) - (R5) about convergent epistemological realism is the following. Its principles - with the exception of thesis (R3) - are, on the whole, acceptable to realists, including supporters of causal internal realism. But they are acceptable as rational-normative principles rather than as factually true historical generalizations - as Laudan interprets them. Thus Laudan is right in his criticism that convergent epistemological realism cannot, judged on the basis of currently available historical examples, be considered strictly true as an empirical theory. But this is compatible with future scientific theories complying with them, of course - should a realist want to defend (R1) - (R5) or their analogues as strictly empirical theses for fully mature science. In any case, our causal internal realism is not as such committed to any of the theses (R1) - (R4) for that matter.

3. However, Laudan of course draws quite different conclusions from his criticisms of convergent epistemological realism. It has not been possible above to do justice to Laudan's rich

discussion because of lack of space. So let me just try to
enrichen our present treatment by quoting Laudan's concluding
remarks and by commenting on their tenability.

Laudan (1984), pp. 136 - 137, claims that his discussion of
convergent epistemological realism warrants the following conclusions:

1. "The fact that a theory's central terms refer does not
entail that it will be successful, and a theory's success
is no warrant for the claim that all or most of its
central terms refer.

2. The notion of approximate truth is presently too vague
to permit one to just whether a theory consisting entirely of approximately true laws would be empirically successful. What is clear is that a theory may be empirically successful even if it is not approximately true.
"Inference to the best explanation" is just a form of
epistemic sleight of hand.

3. Realists have no explanation whatever for the fact
that many theories that are not approximately true and
whose theoretical terms seemingly do not refer are none-
theless often successful.

4. The convergentist's assertion that scientists in a mature discipline usually preserve, or seek to preserve,
the laws and mechanisms of earlier theories in later ones
has not been established and is probably false; his
assertion that, when such laws are preserved in a successful successor, we can explain the success of the
former by virtue of the truthlikeness of the preserved
laws and mechanisms, suffers from all the defects noted
above confronting approximate truth.

5. Even if it could be shown that referring theories and
approximately true theories would be successful, the
realists' argument that successful theories are approximately true and genuinely referential takes for granted
precisely what the nonrealist denies: namely, that explanatory success betokens truth.

6. It is not clear that acceptable theories either do or
should explain why their predecessors succeeded or
failed. If a theory is better supported than its rivals
and predecessors, then it is not epistemically decisive
whether it explains why its rivals worked.

7. If a theory has once been falsified, it is unreasonable to expect that a successor should retain either all of its content or its confirmed consequences or its theoretical mechanisms.
8. Nowhere has the realist established - except by fiat - that non-realist epistemologists lack the resources to explain the success of science."

My quick answers to these claims go as follows. Point 1. is well taken and acceptable to a causal internal realist. As to 2., it should first be agreed that we are still in search of a comprehensible and fully viable account of verisimilitude - but this not much of a criticism against realism. What is clear, however, is that, contra Laudan, a theory cannot be fully empirically successful even when it is not even approximately true. (See our (WR1), (WR2), and (WR3) and the discussion related to them in Chapters 4 and 5.) Laudan operates with a very weak - far too weak - notion of empirical success and so manages to create the impression that his historical examples support his claim.

Concerning point 3., the realist's explanation of course is that some amount of empirical success may of course be achieved a) by coincidence and b) by means of radically false theories which have empirically successful parts or components.

As we have not accepted convergentism above, I will not comment on 4. but go on to 5. Let me just here say that full explanatory success betokens truth in the sense of claim (14) above. To deny that best explanation betokens truth is to deny (14) - and shifts the burden to the non-realist, who has got to refute our argumentation for the truth of (14).

As to 6. I have above defended its thesis on explanation in the sense of our schema (2) and as a rational-normative principle. While I take (2) to be a reasonable schema for scientists to accept let me just here retort - for the lack of space - that its contained principle of explanation could be rejected consistently with the acceptance of causal internal realism. As to Laudan's point 8., let me answer briefly just by pointing to the arguments (WR1), (WR2), and (WR3) for realism (in Chapters 4 and 5), as they just build on the alleged explanatory failure of non-realism.

Laudan ends his criticism of realism with paragraph which I find worth quoting in full:

"To put it specifically in the language of earlier chapters of this essay, the realist offers us a set of aims
for science with these features: (1) we do not know how
to achieve them (since there is no methodology for war-
ranting the truthlikeness of universal claims); (2) we
could not recognize ourselves as having achieved those
aims even if, mysteriously, we had managed to achieve
them (since the realist offers no epistemic, as opposed
to semantic, tokens of truthlikeness); (3) we cannot even
tell whether we are moving closer to achieving them
(since we generally cannot tell for any two theories
which one is closer to the truth); and (4) many of the
most successful theories in the history of science (e.g.,
aether theories) have failed to exemplify them. In my
view, any one of these failings would be sufficient to
raise grave doubts about the realist's proposed axiology
and methodology for science. Taken together, they seem to
constitute as damning an indictment of a set of cognitive
values as we find anywhere in the historical record.
Major epistemologies of the past (e.g., classical empiricism, inductivism, instrumentalism, pragmatism, infalli-
bilism, positivism) have been abandoned on grounds far
flimsier than these."

Laudan certainly puts his claims very bluntly but, in view
of what has been argued earlier in this book, they do not apply
to our causal internal realism. As to (1), our schema (2) - together with what we above said about estimating truthlikeness
and what we will in the next chapter say about the self-corrective nature (and other features) of the scientific method - does
give a general account of how to go about to achieve best-explaining and true theories. However, it is equally important to
see that there is no recursive way for achieving truth, nor
should then such methods be demanded. (Using the theory of
distributive normal forms as above in Section IV, we can say, in
the case of first-order theories, that as there is no decision
method for first-order logic there is no recursive way for
eliminating inconsistent constituents in the polyadic case. Thus
there is no recursive way for defining the probabilities needed,
e.g., for formulas (18) and (23).)
    Laudan's points (2) and (3) have in effect been answered
earlier when discussing (23), with its asymptotic character, and

when commenting on its relation to schema (2) - and note that
finite epistemic accessibility of knowledge of truthlikeness
would be an unrealistic demand also for non-realistic methodol-
ogies (cf. the idealized nature of e.g. warranted assertability). And Laudan's point (4) was already answered above. In all,
Laudan's conclusion about the status of realism - at least when
realism is construed as causal internal realism - could not be
much further from the truth.

In the above discussion we have dealt with growth of scientific knowledge both as a historical and as a normative-rational
phenomenon. There is still a further angle. Scientific method
itself can be said to include the feature of growth and progress, or at least (rational) change. It is often claimed that
the scientific method corrects both itself and the knowledge it
produces, and thus produces new knowledge. This no doubt is
true, and such general aspects connected with the scientific
method in part render support growth theses in schema (2).
However, since these features associated with the scientific
method are rather general and since they also connected with
some other central issues dealt with in this book, we shall
return to them in Chapter 10.

CHAPTER 10

SCIENCE, PRESCIENCE, AND PSEUDOSCIENCE

I. The Method of Science

1. We have in this book repeatedly stressed two central philosophical themes. The first of these is that philosophical,
scientific, and everyday thinking must be free from immutable a
priori transcendental principles. In the context of philosophy
this entails - or at least is compatible with - an anthropological view of man as a part of nature and society (recall Chapter
1). We explicated this partly in terms of the denial of the Myth
of the Given in Chapter 3 and in terms of the adoption of internal or "viewpoint-dependent" scientific realism in Chapter 6.
Accordingly it can be claimed generally that there are no such
things as the World, Truth, Value, Man, Reason, God, etc., using
capital letters here to refer to an a priori privileged view
concerning the named matter. But recall that the other side of
the coin of the rejection of the Myth of the Given is that there
can be no knowledge without prior knowledge, without presuppositions. Accordingly, also a kind of transcendental knowledge,
viz, knowledge of knowledge of objects or "metaknowledge", is
needed, although not in the Kantian sense.

The second leading theme of this book is the importance of
the method of science. This method is the best (most reliable,
valid, and best-explaining, etc.) method for gaining factual
knowledge about the world, we have claimed. We have accordingly
supported scientific realism which makes the scientific method
even the criterion for existence - scientia mensura. But a
critic may now claim that this in fact means giving the method
of science an a priori privileged status in something like the
criticized sense and also the acceptance of scientism in some
pejorative sense (recall the comments in Chapter 7). But that is
not the case, for the method of science is not an immutable
given but a plastic collection of ideas and rules representing
the activities of self-reflective reason.

It should be strongly emphasized that one of the central

features of science, contra e.g. pseudoscience, magic and religion, is its self-correctiveness. Science is plastic and progressive in the sense that it corrects its results (its
"truths") - and, what is more, also its methods. What is most
important in science is accordingly just its self-corrective and
progressive method rather than its substance or content at any
given time.

In this connection it may be noted that while the **scientia
mensura** -thesis emphasizes science in both the substantial and
the methodological sense, the adoption of the scientific world
view and scientific realism does not entail scientism in any
ideological sense. The view adopted in this book is thus compatible with, e.g., "soft" and "green" values.

2. Let us now take a somewhat closer look at science. We shall
start with the general features of the scientific method. How
can we distinguish between what is science and what is not? This
problem - the so-called demarcation problem of science - has
been widely discussed in philosophy. While believing that no
metaphysical solutions to this problem are possible we can still
try to find some methodological standards for scientific activity and the scientific method such that these standards satisfactorily characterize what it is to be scientific. If we succeed in constructing such an "ideal type" of science we can hope
to build out of scientific elements a world view which excludes
pseudoscientific ingredients (such as voodoo and astrology).
Pseudoscience and magic are curses of our time, curses that are
supported by irresponsible authorities, by fear of various kinds
of catastrophies and by human gullibility. (We shall return to
pseudoscience later.)

Science is an extremely complex phenomenon. We shall explicate it further later on, and the following should suffice here.
When we speak of science we can emphasize science as an institution (viz. as organized scientific communities), as a research
process, as a method, or as a body of knowledge. Scientific communities form the part of society which produces knowledge
through research. Research in its turn can be looked upon as
social activity which aims at the systematic and organized pursuit of new knowledge and which obeys scientific methods (cf.
schema (2) of Chapter 9). Such methods are the means of produc-
tion of scientific knowledge, as accepted by the scientific

community. Scientific knowledge, finally, consists of the products of research which have been obtained by these means.

The general characterizations in the above paragraph are
perhaps too circular to be very satisfactory for the understanding of the nature of science. For this reason it is desirable to
attempt to characterize scientific research activity in a non-
circular (or at least less circular) fashion. Several such
features have been described in professional literature - these
characterize above all the scientific method. These partly over-
lapping criteria or characteristics include **objectivity, criticalness, autonomy,** and **progress.** The list could be continued,
and we will continue it later. The above ones are nevertheless
very central and they all (except perhaps autonomy) also serve
to characterize the rationality (at least means-end rationality)
of science. I shall briefly elucidate them but without giving
detailed justifications.

The criterion of the objectivity of science contains, first
of all, the requirement of the objectivity of the domain of
study: science examines real things, be they stones, animals,
electrons or historical documents. Secondly, science is objective at least in the sense of intersubjectivity. The agents of
scientific inquiry are the members of scientific communities and
the motivational background of inquiry is formed by their we-
attitudes (we-intentions, -wants, and -beliefs), not the sometimes idiosyncratic wishes and conceptions of the individual
investigators which enter into the process. Scientific research
processes must at least in principle be public throughout. This
feature also includes (at least in principle) the requirement of
repeatability: e.g., the results of scientific experiments must
be reproducible. (The exact content of the repeatability requirement may be debated, but some relevant - perhaps subject
matterdependent - repeatability should be insisted on.)

The critical nature of science incorporates, not so much
freedom from all presuppositions (for this is impossible, cf.
the Myth of the Given) but a critical and sceptical attitude
towards them. In science nothing is immune to criticism - presuppositions, concepts, theories and hypotheses, theoretical
inferences, experimental set-ups and the performances of experiments, conclusions drawn from data, etc. are all subject to
examination. Of course all criticism and doubt is also based on
its own background assumptions, but they can be varied and
changed. The old simile of the scientific enterprise (philosophy

included) as a ship which is being rebuilt plank by plank while at sea, is still appropriate.

There is one outstanding feature associated with the critical nature of science which is especially central, viz. testability. By testability I mean, more particularly, the empirical testability of scientific theories (the nature of which depends on the field of inquiry). The scientific method is liberal in the extreme with respect to the ideas which enter into theory formation. It is reasonable to allow the presentation of bold and even unlikely ideas. It is important not to suppress scientific creativity - rather, all flowers are allowed to blossom. It is a separate matter altogether that theories and hypotheses are to be testable (and falsifiable), and this requirement must be strictly imposed. Testability itself need be but indirect - i.e. based on (possibly) a great variety of auxiliary hypotheses. But the stricter the tests allowed by a theory, the better. If science does not fulfil the requirement of testability it does not reproduce and develop but stiffens and turns into pseudoscience.

Criticalness and testability are closely connected to self-correctiveness - a central feature of the scientific method not possessed by any other method for gathering knowledge about the world. Critical scientific discussion, which also concerns the results of testing scientific hypotheses, can be expected to lead to a process in which inquiry corrects its own mistakes - be these mistakes in erroneous data or false theories, or even errors in the scientific method itself. In terms of the cybernetic diagram of subsection 8.III.7 we may say that self-correctiveness is based on the cybernetic feedback loops related to the elements x_A^{in} and s_A.

The self-correcting nature of science was defended with great vigor by C.S. Peirce already in the 19th Century (cf. Rescher, 1978). More particularly, he thought that science can systematically eliminate errors. Thus it can be shown for some central scientific tests at least that, when the observations accumulate, an erroneous hypothesis is very likely to be rejected. The same can be said of scientific discussion and debate at large. Unfortunately the matter has not been very thoroughly studied in its general form in the literature (cf. Laudan, 1973).[1] Thus we still lack a unified and precise theory of self-correction, despite the fact that the matter has been examined rather precisely in the case of inductive methods, and despite

the fact that cybernetics and theories of learning offer various
kinds of feedback mechanisms on which a theory could be based.
Since self-correction can apply both to truth and the methods of
obtaining truths, such a unified theory must deal with both
aspects (cf. Rescher, 1977).

Self-correctiveness is a very clear **rationality** feature,
for it is of course rational to be able to correct one's errors
on one's way to best-explaining and true theories about the
world. We can accordingly say that a method for gathering knowledge is the more rational the more conducive to truth it is.

One feature associated with the self-correcting nature of
science seems nevertheless rather obvious: if science cannot
correct its own results and methods, nothing can. It cannot be
done by God, the King, the Party or the oracle of Delphi. In
this respect science is in fact a self-sufficient institution
which does not allow external checks on validity; there are no
extra-scientific criteria of correctness (a fact which does not
rule out several kinds of connections, factual and other, between institutionalized science and the rest of the society).

There are plenty of sad examples of the intervention of,
e.g., political and religious powers in the internal affairs of
science. Indeed, science is self-corrective partly because it is
autonomous. We can, I submit, accordingly accept the principle
of the autonomy of science in the sense that science is or
should be autonomous in the choice of the criteria of truth and
correctness, as well as in the application of these criteria in
scientific inquiry. On the other hand, science can recover from
at least minor violations of its autonomy - as we can judge from
history (cf., e.g., the Lysenko affair). Put in the cybernetic
terms of Chapter 8 the size of the domain of stability, D_T, of a
growth function measures the resistance of science against violations of autonomy. This domain is in general non-minimal, but
it is of course hard to specify its exact size.

Progressiveness was also mentioned as one of the characteristics of the scientific method. As argued above in Chapter 9,
progressive scientific change is possible if the scientific
method is applied rationally. Thus, in reference to our discussion of schema (2), we can say that given that scientists are
in certain specific ways rational (e.g. in replacing their
theories with new, better-explaining ones) science will indeed
grow towards increasingly better-explaining and more truthlike
theories about the world. (Note, too, that the idea of correct-

ive explanation involved in schema **(2)** in fact partly explicates the important feature of self-correctiveness.)

II. Science and Prescience

1. In what follows we shall take a still somewhat more detailed look at science, one which is not confined to the scientific method. Relying mostly on Bunge's (1983) characterization we can look upon science as a cognitive field. We shall say that a cognitive field is a field of human activity in which the aim is to obtain information of a particular area and, in one way or other, to make use of such information. Science is one such cognitive field and religion another, to mention a couple of examples.

A **cognitive field, CF,** can be regarded as an ordered ten-tuple such that **CF** = ⟨A,Y,F,E,D,S,P,T,G,M⟩, where the elements in **CF** (which may be understood mathematically as sets, allowing empty ones) have the following content:

(1)   A represents a community of agents;
(2)   Y represents the host society of A;
(3)   F represents the community A's general (philosophical) views or world view;
(4)   E represents the exact (logical and mathematical) conceptual tools employed by A;
(5)   D represents the domain of inquiry of A;
(6)   S represents the specific knowledge and background assumptions by A which has been obtained from other fields of inquiry;
(7)   P refers to the set of problems of A;
(8)   T refers to the specific pool of information obtained by A through its action;
(9)   G refers to the relevant aims of A;
(10)   M refers to the methods employed by A.

As such the components of **CF** make no reference to either actions or the practical inferences which justify actions. However, they come into the picture in an indirect way, for the community A of course partly gathers its pool of information about the world. This means that A engages in suitable research activity and relies on practical inferences which in fact contain elements from all the components of **CF**.

We can now define a (particular) **science** such as biochemistry ideally as a cognitive field **CF** which satisfies at least the following conditions:

(i)   All elements in **CF** can undergo some changes; the crucial matter is that (3) - (10) may be altered as a result of research results obtained in neighboring fields (cf. E and S in particular).

(ii)   The members of A are rational and well-enough trained to be able to perform relevant practical inferences and to act accordingly.

(iii)   The society Y, or the societies in which the said particular science is practiced, offers the community A autonomy of inquiry and the resources required by its scientific activities.

(iv)   The general philosophical views F contain a) an ontological view of relevant objectively existing real things which can change and act as causal agents, b) an adequate view of the scientific method and c) a view of science as organized activity which aims at least in part at factually truthlike descriptions and explanations, as well as d) ethical rules for conducting research, especially the ethos of the free search for truth, depth, and systematization.

(v)   The exact conceptual tools E consist of up-to-date logical and mathematical theories which can be used, e.g., to sharpen theory formation and to process data.

(vi)   The object domain D consists of real objects, past, present and future (cf. (iv)a)).

(vii)   Specific background knowledge S consists of up-to-date and sufficiently well-confirmed data, hypotheses and theories which have been obtained from the neighboring fields relevant to **CF**.

(viii) The set of problems P comprises, at least primarily, cognitive problems about the domain D and other components of **CF**.

(ix)   The specific pool of information T consists of up-to-date, testworthy and testable (and in part tested and confirmed) theories, hypotheses and data which are compatible with F, as well as of specific information previously incorporated into **CF**.

(x)   The goals and aims G have to do, first and foremost, with
        the search for and application of the laws and theories
        about the domain D, with the systematization and general-
        ization of the information about D into, and for, the-
        ories, as well as with the improvement of the methods M.

(xi)   M consists of appropriate scientific methods which are
        subject to criticism, test, correction, and justifica-
        tion.

(xii)   **CF** is connected with a wider cognitive domain **CF'** whose
        investigators are similarly capable of scientific infer-
        ence, action and discussion as the members of A are, and
        whose members have a similar supporting society (or sys-
        tem of such societies), and for which it also holds that
        a) the components F,E,S,T,G,M of **CF** are in part identical
        with the corresponding components of **CF'** (i.e. they have
        a set-theoretically non-empty intersection) and b) the
        domain D' of **CF'** includes D or else every element of D is
        a component of some system of D'.

The above characterization, which to a great extent relies
on Bunge's (1983) proposal, contains plenty of elements which
cannot be subjected to further scrutiny here. Nevertheless we
can say, quite generally, that it fits well with the thoughts
presented elsewhere in this book. We can see this by examining
the conditions (i) - (xii), one at a time.

Science, we have observed, is creative social activity
which contains no a priori given or privileged elements. This
supports condition (i). Conditions (ii) and (iii) can be held
true even on the basis of what was said in Chapters 8 and 9 as
well as above. These two conditions are nontrivial - not every
society is capable of creating and supporting a scientific
community (cf. theocratic societies). It is equally obvious that
science presupposes philosophical background assumptions (cf.
the Myth of the Given).

Whether all assumptions built into (iv) (especially c)) are
required is perhaps controversial - at least for an instrumentalist. In any case scientific realism satisfies condition (iv)
full well. But as I have shown in above in Chapters 2, 6, and 9,
(iv) can be supported also without direct reference to realism.
But what is more, it can be plausibly argued along the lines of
Chapter 4 that the method of science simply and bluntly is real-
ism-dependent (see also Boyd, 1981, on this). Thus, for in-

stance, science is dependent on inference, in the right circumstances, to best explanations involving reference to unobservables, it may be argued.

Condition (v) needs no deep justification. It can hardly be
denied that muddled thought is far from scientific thinking.
(This condition itself is insufficient to guarantee clear thinking, nor can it without help from the other conditions safeguard
from quasi-precision.)

Condition (vi) can be grounded on the requirement of objectivity (cf. above) and on the conceptual and philosophical
muddles associated with other than the ontology of real objects
(cf. Chapter 7).

Conditions (vii) - (xi) are standard assumptions which can
be accepted without further ado - they harbor no special contro-
versies. Naturally these conditions in particular satisfy the
general characteristics of the scientific method enumerated
above (objectivity, criticalness, testability, self-correctiveness, autonomy and progress). Condition (xii) points to the
systemic character of science - it must dovetail with some wider
cognitive setting. This requirement alone does not rule out e.g.
scientific revolutions.

Scientific activity is both end-rational and means-rational
when pursued according to (i) - (xii). The gaining of scientific
knowledge enhances man's chances for survival - that is a consequence of the successful pursuit of truth (this expresses **end-
rationality**) according to the methods of science (**means-ra-
tionality**). To what extent man is capable of realizing those
chances depends on the success of his practical activities and
their rationality (**practical rationality**).

Our above characterization of science may seem too strict,
and it undoubtedly is somewhat idealized. If we consider the
social sciences and the humanities we can see that at least
conditions (v), (vii), (ix), and (xii) typically are difficult
to satisfy. Still we may regard these conditions as acceptable
ideals in science. We may, however, want to speak of fields of
research which are scientific in a weaker sense than above. Let
us briefly consider this matter.

2. Above we defined the notion of cognitive field as our starting point and went on to characterize science on the basis of
it. But we can indeed define interesting intermediate notions

(cf. Bunge, 1983). Let us first make a broad distinction between **research fields** and **belief systems** and take them to be (at least ideally) mutually exclusive (but perhaps not jointly exhaustive) subfields of the whole family of cognitive fields.

A research field can be viewed as a cognitive field which is based on research and which changes on the basis of research. It is a minimum requirement of a research field that it satisfies, mutatis mutandis, conditions (i), (ii), (iii), (viii), (xi), (xii) of the defining characteristics of science. Condition (i) is of course central and so obvious that it needs no further defense here. Conditions (ii) and (iii) should also be obvious once they have been reformulated so as not to make direct reference to science (but only to research). In (xi) we now speak of research methodology or methodics only. Research methodics is a notion which need not involve the method of science in its full sense. What exactly it should be taken to involve is a broad and difficult topic which we cannot discuss here. As to (xii), it must now be reformulated to speak only of research instead of scientific research in the fullest sense.

A belief field (or faith field, if you prefer) is a cognitive field which either does not change at all or changes due to factors other than research (such as economic interest, political or religious pressure, or brute violence). Thus in the prototypic sense a belief system has nothing to do with research and therefore it negates all the defining characteristics of both science and research field. For our present purposes it is not central to give a more detailed analysis of belief fields here.

Among research fields we may count the basic natural sciences, applied sciences, formal sciences (logic, mathematics), technology (including medicine), jurisprudence, the social sciences and the humanities. Among belief fields we include especially religious and political ideologies as well as various pseudodoctrines.

There are various intermediate notions between research field and science. Thus we may call a research field a **prescience** or **protoscience** if it is a field on its way to becoming a science. More exactly, we may take a protoscience to be a research field satisfying at least conditions (i), (ii), (iii), (iv), (vi), (viii), (x), (xi), and (xii). In addition, there should be at least some reason to think that a protoscience can develop so as to ultimately satisfy the rest of the defining characteristics of science. On the whole, the social sciences can be regarded as protosciences in the defined sense. It should be emphasized that while protoscience is strictly speaking still non-science it should be clearly distinguished from permanently non-scientific fields such as belief fields (as defined above) and pseudoscience. Let us say here in a preliminary way that a pseudoscience is a cognitive field (often but not always a belief field) which is non-scientific (and typically permanently so) but which its proponents still advocate as science. (There are also pseudo-doctrines which are clearly belief systems and whose proponents do not regard them as scientific but claim the opposite, rather.)

It is appropriate to emphasize in this connection that our classifications represent a kind of ideal types. Note in particular that within a science there may exist non-scientific schools, etc. For some purposes at least it might be better to speak of scientific versus non-scientific attitudes, thinking, and action rather than apply these attributes directly to cognitive fields.

III. Magic and Religion

1. We have in this book, and in this chapter especially, given plenty of space for the examination of science and scientific thinking. It may therefore be appropriate to devote a few pages to non-scientific thought. Scientific thinking is, after all, historically a rather new phenomenon, and it has spread slowly. I have no figures to present, but non-scientific thinking still is extremely common - to the extent that many people still attempt, schizophrenically, to think scientifically in some areas and non-scientifically in some other areas (e.g. in the area of religious belief). It is a rather common mistake in ordinary thinking to separate between issues of "knowledge" and "faith", and to imagine that the latter belong to some "spiritual world". It is part and parcel of this mistake to think that questions of faith are somehow quite different from the questions that concern laymen - that they belong to a hidden world of the clergymen.

I would like to claim that at least pseudoscience - but also science - is easier to understand if it is examined in connection with magical and religious thinking. I shall start by a short survey of the central features of magic and religion.

Magic or (systemic) superstition can be thought to be connected with "supernatural" matters, and the same goes for religion, too. Traditional anthropology divides supernatural beliefs
and rituals into magical and religious ones. According to it,
magic has to do with beliefs about impersonal supernatural
powers which - according to these beliefs - are amenable to
manipulation in suitable ways especially by persons "initiated
to" these questions. (In medieval terms, we are here and below
speaking primarily of so-called **low** magic as contrasted with the
more "spiritual" **high** magic; cf. Russell, 1972.) Religion in its
turn consists of (mostly institutionalized) beliefs about per-
sonified, purely spiritual beings who govern man and nature (or
parts of nature).

True, modern anthropology refrains from drawing such a
sharp line of division. Rather, it thinks that all beliefs
concerning supernatural powers are religious in nature. Magic is
nowadays often looked upon as a form of ritual thought and
action which constitutes part of religion (see e.g. Alcock,
1981). It is not very important for our concerns how precisely
the relationship between magic and religion is understood.

Frazer (1922), in his classic study, distinguishes between
two central forms of magic, **homeopathic magic** and **contagious
magic.** Homeopathic magic is based on imitation, and it relies on
a kind of "law of similarity". Magical actions are directed
towards or connected with a thing which in an appropriate way
resembles the object to be influenced. The law of similarity is
then thought to guarantee the desired impact on the intended
object, although it should be plain enough that no such **general**
law of nature exists (scientific research can of course find out
some specific connections of that type).

Voodoo is a typical example of homeopathic magic. Its per-
formers think that sticking pins through a doll which resembles
some person will cause pain in that person. A more homely example is provided by homeopathic medicine which is based on the
assumption that **similia similibus curantur.** Apparently it was
not uncommon in the Finland of the 19th Century to treat hepati-
tis with extract from buttercup (Ranunculus acer), the poisonous
flower with a yellow blossom. This example and other similar
ones indicate that homeopathic medicine is at bottom magic and
pseudoscience. Another example of homeopathic magic is the
search of metals, oil or, say, "earth radiation" by means of a

metal ring (or the like). As far as I know there is no law of nature or other scientifically warranted connection on which the belief in the method could be based - the method survives because of magic, not knowledge.

Frazer's contagious magic, in its turn, presupposes a kind of effective "law of contact": things which have previously been connected with one another are thought to retain some of this connection even when separated. A witch is imagined to be able to urge evil spirits to attack a victim by doing something to pieces of hair, clothing, etc., obtained from the victim. Another example of contagious magic is provided by modern "clairvoyants" who search for missing persons with the help of objects which have belonged to these persons. Such activities have not proved to be successful (cf. e.g. Alcock, 1981 analysis of Hurkos's work). There is no scientific basis for contagious magic of any kind.

There are also other forms and examples of magic. Reading horoscopes, crossing fingers, spitting after seeing a black cat, and fearing the number 13 are just some rather innocent modern examples, taken seriously by rather few people. (I recall having been surprised about the missing floor number 13 in an American hotel.)

All forms of magical thinking and action have a common feature. At bottom they all involve erroneous assumptions about the existence of supernatural beings and powers, and about putative laws which govern them. It may perhaps be difficult to give an exhaustive psychological or other factual explanation of the origins of magic and - equally importantly - of its continued and widespread grip of people in the modern world. One psychological principle which easily comes to mind as a potential partial explanation is operant conditioning. According to this well-known principle of learning a reward which follows spontaneous behavior reinforces that behavior. Skinner investigated the effect in detail, and he observed that most surprising - often rather accidental - behavioral activities can be so reinforced and turned into permanent behavioral patterns. It is hard to say to what extent magical activities are based (consciously or not) on something like operant conditioning. We do know stories of persons who, when succeeding in something, have accidentally worn a particular garment (or the like), and who have since then always worn that garment on similar occasions. (There is a story of a pilot who was conditioned to wear his worn-out pullover during all his war missions, once he survived his first flight wearing that pullover. Analogous stories about athletes are often told.)

Although magical thinking and action can sometimes be a matter of lack of knowledge, this explanation is by no means valid for all cases (although people either cannot or do not want to admit it). Whatever the correct explanation is, such a magical or partly magical attitude ("Spirits just might have an impact on these matters, so I'll do this - after all it's no harm taking precautions") anyway is an enemy of scientific thinking. (On the other hand we must admit that there often is a gap between magical beliefs and magical action - for instance, few people would be inclined to choose their career on the basis of a horoscope.) It should be noted that magic is indeed some-times partly based on real, experienced connections - and this may baffle people. The mistake here is in the inference to a corresponding universal generalization which, moreover, comes to be held a priori (or quasi-a priori) true, i.e. true in an immutable and transcendental sense (cf. the Myth of the Given).

2. Religion typically has to do with beliefs which concern supernatural beings, and with organized human behavior which is in some relevant sense connected with those beliefs. In many religions gods require that men act in certain ways. This gives rise to an ethical dimension which is not always tied up with magic. Otherwise magical and religious thinking and action do resemble one another - especially when they are contrasted with science.

Religion (and in part also magic) is generally based on similar psychological needs (and, no doubt, principles of learn-ing). True, there can be several such effective needs, and the entire matter can be strongly context-dependent. For instance, Freud stressed that the belief in God has its origins in a child's need for "aid and comfort". God can according to this be some kind of substitute for father. Jung emphasized the human need to see life as meaningful and purposeful, and this can provide the foundation for the postulation of a supernatural spiritual being. A great many theoreticians have laid stress on the feature of the belief in God that it diminishes fear and un-certainty. This is likely to be a remarkable explanatory factor, but not an exhaustive one (cf. Alcock, 1981). Whatever the best psycho-sociological explanation for the origins of religion and for its continued existence throughout history is, the phenomenon itself is historically, socially and psychologically extremely important. Insofar as cultural anthropology has any universal generalizations to offer, the generalization that all cultures have (organized) religions is a good candidate.

We are not here primarily interested in the forces which sustain religions, nor in questions of religious convictions. Rather, we want to deal with the relationship between religion and science. Let us briefly examine some beliefs which are central to our culture, viz. the beliefs that there is a divine agent and that this agent actually causally influences worldly affairs. Let us assume that we have some scientific theory (e.g., the theory of evolution), and a rival "theory" based on a divine agent (e.g., the creationism of Christianity). Can we now, within the bounds of this Fragestellung, regard the former theory as scientific and part of the realm of "knowledge", and the latter as unscientific and connected with "faith"? To be able to answer this question we must first examine the notion of a god, viz. agent-god. I shall give a rather rough account and not deal with, e.g., the notion of sacredness or other fine points and distinctions of theology.

A personal god, as it is understood in e.g. Christianity or Judaism, can be defined as follows (cf. Swinburne, 1979, Mackie, 1982):

(A)   A (personal) god is an agent which has no body (and which is pure "spirit"), which is eternal, fully free, omnipotent and omniscient, fully good, and the creator of all things.

Such a transcendent notion of a personal god can in fact be claimed to be logico-conceptually incoherent. (For what sense can be made, e.g., of states of knowledge, willings, actions, etc., attributed to something without a body?) However, let us here suppose, for the sake of argument, that this is not the case. A doctrine which is based on a personal god naturally assumes that there is a god. Such an existential claim (and consequently the whole doctrine) can be interpreted either literally or metaphorically. The former interpretation - as I will understand it, at least - is ontologically significant, that is, according to it god exists as a real being with causal powers.

(The concept of a real being can at least in part be defined on the basis of causal influence, as we already remarked in Chapter 6.) In what follows I shall only deal with the ontologically significant idea of creation which involves causal influence upon the world, no matter whether this takes place directly or somehow indirectly via "ordinary" real beings and phenomena.) A metaphorical interpretation of god-talk can be ontologically significant, but it can equally be ontologically innocuous, in which case it is of course not thought to involve ontological commitments. (This latter notion applies to Kant's "regulative" concept of God, to give an example.)

We can accordingly include within the realm of matters of faith, first, the non-ontological interpretation of the doctrine of divine agent. Such a doctrine cannot contradict science, for it does not make ontological claims about the world. How about the second or ontological interpretation? For creationism to be even a prima facie rival for the theory of evolution it must be interpreted ontologically. But then we must note that a doctrine which relies on God as a causally efficient agent can not be a proper (rival) scientific theory. This conclusion can be supported in several different ways.

One of these ways is this. The mentioned doctrine is one which is not testable and which is able to give only ad hoc explanations (cf. "Event A happened because God willed it" vs. "Event A happened and God only let it happen without willing it to happen"), at least if the details of the mechanisms through which this agent influences the real world are not described in a testable way - and it is a well-known feature of religion that it ordinarily does not and cannot describe them (because, e.g., of the assumptions of full freedom, omnipotence, and omniscience and the vagueness of god-talk). In those rare cases when some such informative and testable claims have been stated, e.g., that the earth was created a few thousand years ago, those claims have been blatantly false.

Without going into details we can say that the god-doctrine (e.g., creationism) is most certainly not a scientific cognitive field nor a science in the sense explained above. (The god-doctrine fails to meet at least conditions (i), (iv) - (vii), and (ix) - (xii) which are constitutive of a science.) It is thus clear that the god-doctrine can not in principle, even with modifications, give such a best explanation on which the **scientia mensura** thesis of scientific realism relies. Let us call

this result thesis **(B)**. In this way we can say that definition
(A), thesis **(B)** and the scientia mensura -thesis logically imply
that an agent-god cannot exist as a real, causally efficient
being. (This deductive connection holds on the qualification
that we interpret the scientia mensura thesis in a sense
stronger than minimal scientific realism **(MSR)**, e.g., in the
sense of **(MOSR)**, **(ESR)** or **(SSR)**.)

We can say quite generally that postulating an ontologically efficient god as a factor which allegedly explains phenomena in the real world is, for a scientist, merely to explain a
phenomenon which requires explanation by using a factor which is
still murkier and more in need of explanation. Nor does explaining the "purpose" or significance of life or the acceptance of
Christian morality require such postulation.

On the basis of what has been said it should be clear that
systems of belief such as magic and religion are far from
science and scientific thinking. A scientific world view and an
ontologically significant world view based on religion are
rather different and incompatible matters. This incompatibility
is based both on ontological and especially methodological considerations, as we in fact already observed. Both in magic and
in religions there are principles which have been assigned a
privileged and transcendental status in the sense of the Myth of
the Given. Needless to say this is contrary to the method of
science, for in science there are no absolute givens. It is
equally clear that magic and religion are not, as is science,
self-corrective doctrines.

IV. Pseudoscience

1. The scientific world view is very rare. My guess is that at
least 99% of all currently living human adults have a non-scientific world view and way of thinking. Most people probably base
their lives on religion and/or magic. Not surprisingly, magic
and religion are relevant also for pseudoscience, as will soon
be seen. But before that let me amuse the reader by mentioning
some results a Gallup investigation conducted in the U.S. in
1978 produced. According to it, 57% of all Americans believe in
ufos, 54% in angels, 51% in ESP, 39% in devils, 37% in precognition, 29% in astrology, 24% in clairvoyance, and (only!) 11% in
ghosts (cf. Greenwell, 1980). Irrespective of whether this in-

vestigation was waterproof we can see the general sociological situation in one of the most civilized and educated societies in the whole world. The figures speak for themselves and show both the prevalence of magical and religious thinking as well as its factual connection with pseudoscientific thinking.

To go into more philosophical connections, magic, religion as well as much of pseudoscience deal with an obscure mentalist or "spiritual" ontology (cf. spirits, gods, demons, fairies) and that is true of many pseudosciences as well. We may put this more generally and say that magic, religion and many pseudosciences deal with paranormal entities and phenomena. To say that a phenomenon is paranormal is often understood in the broad sense to mean that it cannot in principle fall within the ontology of science (or, alternatively, at least be explained by science). (A more specific characterization, defended by Braude (1980), goes as follows. A phenomenon P is paranormal just in case a) P is inexplicable by current science, b) P cannot be explained scientifically without major revisions elsewhere in science, and c) P thwarts our familiar expectations about what sorts of things can happen to the sorts of objects involved in P. But here we need not adopt any specific explication of paranormality.)

Next, the mentioned belief fields typically rely on authoritarian epistemology (cf. God's revelation as a source of certain and immutable knowledge) and dubious aprioristic transcendental principles. Inferences are typically based on questionable a priori principles. Thus in typical magical thinking as well as in much pseudoscientific thinking the "principle of similarity" is involved (recall Section III).

2. It is time to go to a more detailed examination of pseudoscience. We said earlier that pseudoscience is a belief system whose supporters, however, incorrectly regard it (knowingly or not) as a science or a branch of science. Can pseudoscience and pseudoscientific thinking (and action) be characterized in more detail? If we take such pseudosciences as astrology, the theory of biorhythms, suitable parts of parapsychology and ufology, homeopathy and faith healing we may arrive at the following view (cf. Bunge, 1983, Radner and Radner, 1982).[2]

If we think of pseudoscience as a cognitive field, **CF**, of the kind **CF** = ⟨A,Y,F,E,D,S,P,T,G,M⟩ as above in Section I, we can see that pseudoscience differs from science with respect to every element of **CF** (recall the defining conditions (i) - (xii) of science). I shall below take up some conspicuous differences. Probably nothing like strict necessary and sufficient conditions of pseudoscientific thinking can be given. We must instead deal with a kind of ideal type here. The following list characterizes it:

(1)   A pseudoscience typically or often relies on a) a very obscure and ill-defined ontology (consisting of, e.g., unembodied spirits - cf. witchcraft, parapsychology) and on b) an epistemology which accepts epistemic justification on the basis of authority or is based on some alleged paranormal epistemic abilities of an elite such as priests, and on c) a dogmatic attitude and ethos to defend a doctrine even by non-scientific means as contrasted with scientific search for truth (cf. creationism, Lysenkoism). Criticism is not welcomed by pseudoscientists. (Recall, on the other hand, the properties (iv), (vi), (x), and (xi) characteristic of science.)

(2)   Pseudoscientific thinking often shuns conceptually and logico-mathematically exact thinking - cf. creationism, witchcraft, psychotherapy. (Recall (v) of the properties of science.)

(3)   The hypotheses and theories - to the extent there are any - of a pseudoscience are generally either impossible to test or badly supported, both empirically and theoretically. This feature is indeed characteristic of all pseudoscience. Pseudoscientists often (incorrectly) try to compensate for qualitatively poor evidence by piling up great quantities of such evidence. Parapsychology provides a typical example of this. (Cf. (vii) and (ix) in the case of science.)

(4)   The hypotheses and theories of pseudoscience are not changed due to confrontation with empirical and other evidence. They often flatly contradict well-confirmed scientific hypotheses, but this fact does not usually have much effect on pseudoscientists, who "know better". On the whole very little change takes place in pseudoscience and such change is typically due to factors not related to research. (Recall parapsychology, Lysenkoism, flatearthism, etc., and contrast with properties (i), (vii), and (ix) of science.)

(5)   Pseudoscience involves anachronistic thinking, viz. thinking going back to old, refuted theories and assumptions; cf. e.g. creationism, flatearthism, and confront this with features (v), (vii), and (ix) characterizing science. For instance, creationists argue against evolution theory by claiming that mutations always involve something harmful to the organism - but this is a false and refuted claim.

(6)   Pseudoscience often appeals to myths (cf. von Däniken on ancient astronauts, creationism) and unfounded mysteries (cf. the Bermuda Triangle, ufos, poltergeists). (Recall characteristics (iv), (v), (vii), (viii), (ix), and (x) of science.)

(7)   The problematics of a pseudoscience often consists in practical rather than cognitive problems (cf. such problems as how to bring about a certain effect, for instance, how to feel better or how to cure a certain illness or disorder). Thus indeed pseudoscience often contains aspects of pseudotechnology. (Cf. features (viii) and (x) of science.)

(8)   The methods of pseudoscience are unscientific especially in the sense that they are not self-corrective and checkable by alternative (especially scientific) methods nor are they based on well-confirmed general theories. This applies to all typical pseudoscientific doctrines. (Cf. characteristic (xi) of science.)

(9)   Pseudoscience is generally a doctrine or body of doctrines isolated and distinct from the science of its time. (Cf. creationism, Lysenkoism, flatearthism and feature (xii) of science.)

Our above characterization of pseudoscience shows that criteria (i), (iv) - (xii) of science do not apply to pseudoscience but rather to their opposites. As to features (ii) and (iii), the members of a pseudoscientific community can be called believers, and typically they lack the education and skills of proper scientists. The societies which support pseudoscience do it mostly through tolerating it rather than actively providing it with resources - unless perhaps big money is at stake or the political or religious system of such a society requires it (cf. Lysenkoism).

Such examples of pseudoscience as the theory of biorhythms, astrology, dianetics , creationism, faith healing may seem too obvious examples of pseudoscience for academic readers. A more exciting case is provided by parapsychology, for it is, indeed, a very controversial field. Let us have a brief look at it.

3. Parapsychology supposedly studies parapsychological phenomena, viz. phenomena which are both psychological and paranormal (recall our above characterization). These are commonly classified as phenomena of telepathy, clairvoyance, precognition, and psychokinesis. The first three of these are jointly called ESP- phenomena ('ESP' standing for 'extrasensory perception'). All parapsychological phenomena are jointly called psi-phenomena (and parapsychology might perhaps accordingly be called psience!). Now it is a very peculiar feature about parapsychological research that it is almost exclusively concerned with showing the existence of psi-phenomena. As far as I know there is no proper science the existence of whose subject matter would be similarly under doubt.

Is there good evidence for parapsychological phenomena, then? If you ask the man in the street (or even a typical parapsychologist) he is likely to say that, for instance, telepathic phenomena exist (recall the Gallup result referred to earlier). In many cases the evidence of the man in the street presumably is anecdotal ("My friend told me that such and such peculiar psi-phenomenon took place about a decade ago"). But such stories typically are unreliable and allow for too many widely differing interpretations. Accordingly, they are almost useless for strictly showing the existence of parapsychological phenomena. The same goes for sensational books and magazines. (Parapsychologists also tend to agree with this judgment.)

So we must look at evidence coming from controlled scientific experiments. To put my neck out immediately, I claim, contrary to the opinion of many parapsychologists, that there still are no repeatable experiments for showing the existence of parapsychological phenomena. That is, there are no recipes for creating conditions under which such phenomena will always or typically take place. In fact, the situation is worse: there are no repeatable experiments even for showing statistically significant deviations (from normality) in people's relevant abilities to behave, defining deviations here with respect to either

"random" situations or "normal" situations - as the case may be.
Thus there are no repeatable experiments indicating the existence of such non-normal abilities, still less is there evidence
for invoking telepathy, clairvoyance, precognition, and psychokinesis as best-explaining factors for such alleged deviating
behaviors.

I have above made strong claims. How can they be substanti-
ated? If you look at parapsychological literature you get the
impression that the existence of parapsychological phenomena has
been clearly proved (see e.g. Wolman, 1977, Grattan-Guinness,
1982, Eysenck and Sargent, 1982). However, critical investigation has shown that such claims are not well taken at all (see
e.g. Alcock, 1981, Marks and Kammann, 1980, Johnson, 1980,
Frazier, 1981, Randi, 1982, Hansel, 1980, Gardner, 1982, Abell
and Singer, 1981, Hyman, 1982). Perhaps the most important
reason for saying so is simply that the experiments have been
badly done, even if ostensibly positive evidence for psi-phenomena seems to have been obtained. Let us consider the matter in
some detail.

For one thing, there has been very much cheating in the
experiments throughout the history of parapsychology (see e.g.
Randi, 1982, Marks and Kammann, 1980). When designing parapsychological experiments one should accordingly take precautions
against magicians' tricks. There are also several other sources
of experimental error. Thus there is the problem of the experimenter effect (e.g. the experimenter's pro-attitude versus
con-attitude towards psi and the "Atlantic effect": significant
results in U.S. experiments have been claimed not to be reproducible in English laboratories, and vice versa). There is the so-
called sheep-goat effect, the decline effect, the "shyness" effect, the psi-missing effect and other error sources (such as
statistical problems). I shall not here take them up (see e.g.
Alcock, 1981, for an expert discussion).

Let me, however, mention Hyman's (1982) recent evaluation
of experimental parapsychology here. Hyman, contrary to such
rather extreme sceptics as e.g. Randi, Gardner, Hansel, and
Alcock, is more sympathetic towards parapsychological research
and certainly wants to give it the benefit of the doubt. He
mentions and discusses three experimental fields in which parapsychologists think best progress has been made. These are the
so-called remote-viewing experiments (especially those by Targ
and Puthoff), Schmidt's random generator experiments as well as

the **Ganzfeld** experiments. But as he notes, heavy criticisms can be directed against all of these studies. As to remote-viewing, see especially Marks and Kammann (1980) and Randi (1982). Schmidt's famous experiments have been criticized by Alcock (1981) and others. Hyman (1982) argues that due to several statistical and other problems the alleged significant results of **Ganzfeld** studies are illusory. So, in all, there is nothing in the experimental results to convince at least a sceptical (but open-minded) observer.

I should here also emphasize the often mentioned feature that there are no proper theories in parapsychology, not even "models" in the sense there are in otherwise comparable fields in experimental psychology. But there cannot be very "hard" data without theories. Unless alleged experimental facts can be backed by theories they often are not well confirmed but rather soft.

What disturbs me much as a philosopher is that parapsychological phenomena, if they existed, would contradict (or at least seem to contradict) well-established scientific theories. Only telepathy at very short distances would seem to be compatible with fundamental physical theories. In other words, the existence of parapsychological phenomena would require a drastic revision in our fundamental scientific theories. And that is a consideration which certainly must affect one's evaluation of the alleged parapsychological facts. (Parapsychologists might want to say here that parapsychology deals with unembodied spirits which are not bound by the laws of physics - but that would be an irresponsible response unless clear conceptual sense is made of such paranormal phenomena, and until their existence is proved in a strong sense satisfying a sceptic.)

Let us finally compare parapsychology with the properties of pseudoscience as stated earlier. Referring to our earlier numbering, much of parapsychological research satisfies the central criteria of pseudoscience. To wit, parapsychology relies on an ill-defined ontology (feature (1)) and typically shuns exact thinking ((2)). The hypotheses and theories of parapsychology certainly are in bad shape (criterion 3)). Extremely little progress has taken place in parapsychology on the whole and, furthermore, parapsychology conflicts with established science ((4)). Parapsychology is also poor in its research problems, being mainly concerned with establishing the existence of its subject matter and having practically no theories to create proper research problems (feature (7)). While in parts of parapsychology there are attempts to use the methods of science there are also unscientific areas and in any case parapsychological research can at best qualify as prescientific because of its poor theoretical foundations (cf. (8)). Furthermore, as emphasized, parapsychology is an isolated research area (cf. (9)).

How about criteria (5) and (6)? Does parapsychology involve anachronistic thinking and rely on myths? Perhaps we must answer negatively in the case of much of experimental parapsychology, but there are anyway more exotic fields like the study of reincarnation, out-of-body experiences, and poltergeists (and other ghostly phenomena) in which many kinds of mythical ideas have been relied on (see Alcock, 1981). Indeed, parapsychology is both historically and systematically connected with magic and religion. Its obscure ontology certainly has much to do with this (cf. ghosts, spirits, demons). Also some of the nonscientific thinking to be found in parapsychology relates to this. Let me mention a couple of examples relevant to experimental parapsychology as well. First, if it is taken for granted (as it is in some psi-circles) that dreams regularly (or often) serve to predict the future, such thinking relies on backward causation (maybe operant conditioning is psychologically involved here!). Or think of homeopathic magic (e.g. voodoo). Accepting the principle of homeopathic influencing clearly entails the acceptance of psychokinesis.

On balance, there are then good grounds for regarding much of parapsychology as pseudoscience. However, I would still not quite join Alcock and many others in regarding all of parapsychology as pseudoscientific - it is partly prescientific, I would rather say.[3] But I think it is reasonable to require strict demonstration of the existence of psi-phenomena in view of (1) the rather incredible history of cheating in parapsychology and (2) the incompatibility of parapsychology with well-established theories and laws of science. And presently we do not have such waterproof demonstrations.

4. One could write a lengthy book on the different pseudosciences. Space does not allow me to attempt to demonstrate the pseudoscientificalness of different disciplines. Let me merely mention some examples of doctrines which either have been con-

vincingly shown to be pseudosciences or which at least are prime candidates. The following short list speaks for itself: astrology, parapsychology (in part), ufology (in part), the three-stage theory of biorhythms, iridology, Lysenkoan biology, creationism, (some) empirical applications of catastrophe theory, anthroposophic agriculture, homeopathic medicine, and at least some forms of psychotherapy.[4]

Pseudoscientific thinking is widespread. It flourishes all-over, even within scientific communities. (Doctrinal edifices are seldom entirely monolithic - science can contain pseudoscientific islands and pseudosciences scientific islands.) Pseudoscience is much more popular and profitable than science. It brings about troubles to scientific policy makers at least when they do not have a sufficient scientific background. Pseudoscience is more of a headache for laymen than for scientists or scholars, for the latter, or scientific communities in general, easily distinguish pseudoscience from science on the grounds that it does not apply the scientific method.

A greater difficulty for science is to distinguish from one another a budding, promising, and possibly unorthodox protoscience, and enticing unorthodox research which eventually turns out to be an unproductive hybrid and possibly a pseudoscience (cf. the so-far unsettled fate of sociobiology). Both may instantiate some sort of application of the scientific method - the rejection of Wegener's theory of continental drifts for half a century is an example of groundless rejection of a protoscience. We have here a case where a scientific community has dogmatically committed itself to normal science. (Presently a modified version of Wegener's theory is part of current normal science in the field.)

We can note that the problems for a layman are magnified. Many times he ought to be able to distinguish between pseudoscientific and scientific inquiry, and within science between unfruitful and really significant research. This problem is especially pressing in areas where science still is at an infant stage or at least has not reached proper theories nor theoretically grounded applications; and this situation, as we know, is not uncommon. As examples we could mention problems of education, of nourishment, as well as the multifarious difficulties having to do with health care and medication.

As to education, we can here note quite generally that many of us (or at least of our parents) have been brought up by

methods which are rather different from the liberal methods we
use. A concrete example: still in the 1930's it was held, em-
phatically, that a child must learn its mother tongue before it
is exposed to another language. The scientific recommendation of
this decade seems to be that children learning two languages at
a time do not suffer from drawbacks of simultaneity (as long as
language instruction is given in the right way) - in fact it is
maintained that all this increases their mental capacities.
There is all reason to wonder what changes future educational
research will bring about!

Nourishment problems, and the related issues of health care
and medication are rather vexing, for also in these matters
action is guided by rather insufficient evidence (after all, it
is rather difficult to carry out double-blind experiments on
human beings which last for decades). Similarly, there are no
comprehensive but yet sufficiently specific theories in the
field. It follows that at least a layman does not run out of
questions. Who can give a scientifically grounded answer to the
question if I or you should eat mixed food, be a vegetarian, or
perhaps a lacto-ovo-vegetarian (to maximize our health)? What
about the disputes of the effects of some specific kinds of
food? Should one avoid food with high cholesterol contents (or
is avoiding such food irrelevant), or should one eat extra
selenium, fluorids, and vitamin C in addition? Can we eat our
tomatoes and cucumbers without peeling them (for fear of beno-
myl), while we already peel our fruit? Should we alter our
habits whenever novel statistical correlation results from field
studies are obtained? Should we rely on "established" medicine
or rather attempt to find our own way or even resort to quacks
if the need seems to arise?

A layman faces such questions in his daily life - whether
he recognizes it or not. Many people escape to pseudoscience
(e.g. to macrobiotics in nutrition issues), the majority probably remain passive (and maintain some kind of status quo).

The important thing about all this is the scientific attitude in contradistinction to a pseudoscientific or magical one.[5]
Pigheaded reliance on present scientific results cannot be considered sensible either. The science which is said to be the
measure of what there is is the science of distant future - not
of today - , but a scientific attitude is a matter of today, not
tomorrow.

NOTES

CHAPTER 1

1)   Kant's notion of a priori knowledge is very strong. As, for
instance, Kripke (1972) and Kitcher (1980b) have argued against
Kant, if something p can be known (by some or all human beings)
a priori it need not be necessary that p (e.g., my knowledge
that I exist might be regarded as a contingent a priori truth).
      Kitcher (1980a) has presented a weaker analysis of a priori
knowledge which seems worth further consideration. His analysis
is based on the general idea that a priori knowledge is knowledge obtained in such a way that it could have been obtained in
the same way if the knower had had different experiences.

2)   Kant's transcendental philosophizing does not - and is not
meant to - cover all traditional metaphysics. Thus it is supposed to be concerned with the a priori limits of human knowledge but not - as proper metaphysics - with anything transcendent, viz. anything going beyond those limits. However, as we
shall argue or at least indicate later in Chapter 6, Kant never
succeeded in drawing a clear distinction between what belongs
to the realm of human knowledge, viz. phenomena, and what goes
beyond it, viz. noumena or things in themselves. This is mainly
because, according to Kant, things in themselves are unknown to
us. But this in turn is because our capacity of knowledge is (at
least partly) unknown to us, it may be argued. Thus the insepar-
ability of meta-knowledge and object-knowledge entails that the
notion of transcendental knowledge (in the sense of meta-knowledge) is intelligible just to the extent the notion of noumenon
or the thing in itself is (cf. Hintikka, 1984).

CHAPTER 2

1)   As will be argued, questions of existence are to be settled

236

by reference to best-explaining science. Thus we need not read
more into the Sellarsian dichotomy "exists" versus "really exists" than that the latter refers to best-explaining science
while the former does not. We might, for instance, construe
existence in the manifest image relative to best-explaining theories within that conceptual framework. However, if our arguments of Chapters 4 and 5 below carry weight that cannot amount
to best-explaining science and hence to a specification of what
"really exists".

2)   The recent book by Stich (1983) contains a good discussion
of the nature of mental states. But contrary to what Stich suggests (on pp. 47 - 48), our present Sellars-type account is not
a "narrow" causal account but rather something that might be
called a "broad" causal account. By a broad causal account of
beliefs, etc., I mean roughly one accounting for (types of)
beliefs in terms of their causal relations to their environment
(understood in a full-blown realist sense), to other beliefs and
other mental states, and to behavior (including behavior and
action in full-blown achievement sense). In addition, the intentional features of beliefs are accounted for in terms of the
analogy theory of thinking by reference to "believings-out-loud"
and other overt speech.

CHAPTER 3

1)   Sellarsian dot-quotation is a functional operation which is
based on meaningful rule-following (verbal and non-verbal) behavior, including intentional action. The rules in question are
a) world-language, b) language-language, and c) language-world
ones. The functional relationships in question can in principle
be stated explicitly so that, for instance, logic can be founded
on them (or rather on a suitable subclass of them) and be represented in terms of them. In this connection reference can be
made to the results of Gärdenfors (1984). He has succeeded in
defining or analyzing the logical connectives as well as logical
truth in terms of the functional relationships holding between a
class, K, of states in terms of mathematical category theory.
However, these results so far concern only propositional logic.
Although he himself interprets K as a class of belief-states, we
may as well take the elements of K to be meaningful behavior-

episodes in the above sense (or, if you prefer, conceptual psychological states, given the analogy account of thinking). Furthermore, the investigations by Gärdenfors should be generalized to cover all the above types a), b), and c) of functional relationships, without essential simplifications.

2)   If we replace the expression "He is a bachelor" in (h) by "·He is a bachelor·", viz. the corresponding dot-quoted expression, we arrive at a more realistic formulation. Switching to the corresponding may-rule instead of the ought-rule used here would make the analysis still more realistic, it seems.

    Even when using such a may-rule in the analysis the relevant causal impact of the world on the speaker is still there, so to speak. For I suggest that it should anyway be regarded as an acceptable explanatory hypothesis, in the case of simple descriptive sentences at least, that if the speaker uttered the sentence in the right or optimal circumstances (as required by the rule) then the utterance was brought about in part of the state of affairs expressed or pictured by the sentence in question.

3)   Sellars has written about his views on semantics and ontology in several contexts (see e.g. Sellars, 1974, 1979, 1981, 1983). The critical writings by Loux (1977, 1978) and Marras (1978) are important in this context, too. Sellars (1979) contains an interesting correspondence between Sellars and Loux on Sellars' semantical views.

CHAPTER 4

1)   To intuit is to represent a "this". Accordingly, Kant's intuitions seem to be analyzable according to a linguistic model, which regards them (or more exactly conceptual intuitions as opposed to non-conceptual ones) as singular (or nongeneral but category -involving) representations of the form 'this-such', e.g., 'this table' or 'this brown table' (cf. Sellars, 1967b). While 'this table' may be regarded as representing a single intuition 'this brown table' stands for an intuition of a manifold (derived, as it were, from "This is a table and it is brown."). (This kind of linguistic way of interpreting conceptual intuitions is not without its problems, however; cf. Thomson, 1972.)

2)   Kant's (1787) own argumentation for his Copernican revolution is very problematic, as is commonly agreed. It can be (and has been) claimed that what is basically at stake in Kant's epistemology is a (partially) failed attempt to go from the traditional "perception-model" of knowledge (knowledge as knowledge of objects) to the modern propositional view of knowledge (knowledge as knowledge of propositions). Kant makes a problematic concept-intuition distinction and the still more problematic assumption that intuitions are "given" to us as a manifold out of which only concepts can help to make a unity. This seems much like an ad hoc -assumption that Kant needs in his transcendental deduction to make his Copernican revolution work and to demonstrate the possibility of synthetic a priori knowledge. It can be claimed that a purely propositional view of knowledge, which builds on predication instead of Kantian synthesization, does not need the above Kantian assumption. (See Rorty, 1970, and 1979b, pp. 148 - 155, for this kind of argumentation, which we need not take a definite stand on here. Compare this with Sellars' reasonable "linguistic" Kant -reconstruction in which conceptual and non-conceptual intuitions are clearly distinguished and given distinct roles - contra Kant.)

3)   The constructive empiricism of van Fraassen has been widely criticized - see, e.g., Musgrave (1984), (1985) and Hooker (1985) as well as other papers in Churchland and Hooker (eds.), (1985).
       To mention one of Musgrave's points, he argues that van Fraassen is not in a position to make a coherent observable - unobservable dichotomy, which, however, his constructive empiricism requires in order not to collapse to full blown realism. For van Fraassen makes it a matter of scientific theory to decide what is observable and what is not. Suppose now a theory T says that something A is observable while B is not. But a constructive empiricist can only accept T as empirically adequate and thus believe to be true only what T says about the observable. But the statement "B is not observable by human beings" cannot without contradiction be a statement about something observable by humans. Thus a consistent constructive empiricist cannot believe any statement of this kind to be true and thus cannot draw a workable observable/unobservable dichotomy at all.

4)   Putnam (1978), p. 20, has recently interpreted realism with
reference to Boyd as a general empirical hypothesis consisting
of the following claims:

(1)   Terms in a mature science typically refer.
(2)   The laws of a theory belonging to a mature science are
       typically approximately true.

Scientific realism so understood is supposed by Putnam to be an
explanatory theory which is meant to explain that scientists act
as they do because they believe (1) and (2). Their strategy
supposedly works because (1) and (2) are true, and thus realism
explains the success of science, we may say. This kind of realism may also be taken to explain that science is convergent,
according to Putnam.

        However, as I have argued elsewhere,   Putnam's claims can
be accepted at best if we understand the notions of reference
and truth in (1) and (2) in a realist way whatever that realist
way is exactly taken to involve (see Tuomela, 1979). Given that,
(1)   and   (2) already presuppose realism if they are able to
function as claimed. Yet, as we will see in Chapter 7, (1) and
(2) can be taken to explain the success of science if reference
and truth are realistically understood and if approximate rather
than   strict reference and truth is meant. (For other related
criticisms against the present Putnamean view of realism see,
e.g., Bradie, 1979, and Laudan, 1981.)

5)   In discussing this aqua regia -example van Fraassen (1980),
p. 33, asks three questions: "Did this postulation of microstructure really have no new consequences for the observable
phenomena? Is there really such a demand upon science that it
must explain - even if the means of explanation bring no gain in
empirical predictions? And thirdly, could a **different** rationale
exist for the use of micro-structure in the development of a
scientific theory in a case like this?"
        To the first question a realist can answer that the example
only requires that there be no observable, non-theoretical factors to account for the variation of dissolution rates. It is of
course a demand of good explanatory theories that they yield
more observational consequences than what they were introduced
to explain. But this does not speak against the realist's pres-

ent argument meant to illustrate and argue for the nomic unstability of the manifest image. Thus he need not answer affirmatively to the second question. As to the third question, the
aqua regia -example just purports to show that the realist
strategy is the best one - and the only leading to true laws and
theories.

6)   We shall in this chapter follow the common practice of
formulating the problems surrounding Bell's inequalities partly
in experimental or quasi-experimental terms, but it is important
to realize that what we in principle are concerned with here is
microsystems ("quantons") with certain spin component values,
independently of how such systems and their properties have been
created and "prepared".

CHAPTER 5

1)   Sellars' (1977) central argument against instrumentalism is
related to just this. For he claims that the so-called standard
account of theories (that you find in, e.g., Hempel's earlier
writings) is false just because there are no stable empirical
generalizations without injections of theory, viz. there are no
"full lawlike observation framework counterparts" (of theoretical theorems) in the absence of theoretical considerations. He
suggests that "an observational counterpart can be a full counterpart only by making use of Ramsey's device, for the observation framework does admit of the introduction of predicate
variables and of quantification over them" (Sellars, 1977, p.
320). But, and this is the crux, "the predicate variables of the
Ramseyized full counterparts cannot, in principle, be cashed out
in terms of perceptible predicates", he claims. His arguments
for this basically hang on the explanatory incompleteness of the
observation framework or, if you like, the manifest image.

2)   If we opt for the greater generality afforded by the reinterpretation account we can substitute the following definition
(IDE^*) for (IDE) and use the notion of counterpart predicate
(viz. member of λ^* below) and sentence in a sense which can be
defined by means of the translation function tr to be discussed
in Chapter 9 (also see Tuomela, 1984, Chapter 14):

(**IDE***) The predicates in μ of a scientific theory $T(\lambda^*\cup\mu)$ are **logically indispensable for deductive explanation** with respect to λ and the explanatory relation ε if and only if

(1) for some (general or singular) sentence F in $L(\lambda)$, $T(\lambda^*\cup\mu)$ deductively explains F^*, the counterpart sentence of F in $L(\lambda^*\cup\mu)$, in the sense of ε-explanation;

(2) there is no sentence $T'(\lambda)$ employing as its extra-logical predicates only those in λ such that the counterpart T'^* of T' in $L(\lambda^*\cup\mu)$ is a subtheory of T and that T'^* also explains F^* (in the sense of ε-explanation); nor is there a subtheory $T''(\lambda^*\cup\mu')$ of $T(\lambda^*\cup\mu)$, $\mu'\subset\mu$, with that same explanatory power.

3) Note that if needed we can conclude to the existence of unobservable objects at hardly any extra cost by the following modification of premise (ii). Add to the end of its antecedent the phrase 'and if they purportedly refer to unobservable entities' and instead of the expression 'and typically unobservable' we may then simply use 'unobservable' in the consequent of (ii). (Premise (iii) should then of course be modified in accordance with this.)

4) To compute the depth of formula (**A**) we formalize it by the following statement:

$$(y_1)(y_2)(y_3)(y_4)(x)$$
$$[S(y_1,y_2,v)\&M(y_3,b,y_1)\&M(y_3,y_4,y_2)\&c(x)=y_3\&r(x)=y_4].$$

Here

$$S(y_1,y_2,v) \text{ if and only if } y_1+y_2=v$$
$$M(y_3,b,y_1) \text{ if and only if } y_3\cdot b=y_1$$

Formula (**L**) is treated analogously.

5) Recall that we are presently making the simplified and idealized assumption about the existence of such correspondence rules. In subsection 2 below we shall indicate how to remove this simplification in favor of the more realistic (in a double sense!) reinterpretation approach.

CHAPTER 6

1)   Note, by the way, that if one accepts a Sellars-type semantics (where semantic statements like 'X refers to Y' are intra-
linguistic) there are still going to be something like **Ersatz**-
reference relations, even if reference proper is an intralin-
guistic matter and not a relation at all (cf. Ch. 3 and below).
These **Ersatz**-references are of course the relata that Sellarsian
world-language (and language-world) uniformities involve. Recall
that these uniformities are complex non-conceptual psycho-socio-
historical uniformities specified by the theory which best explains man's complex connections with his environment.)

2)   Let me here still mention another flaw in Putnam's (1981)
discussion. He claims that such notions as reference and explanation are irreducible to materialistic (or physicalistic) notions. I will not discuss his arguments but only say that I
agree with his claim (see Chapter 7 below). However, he goes on
to claim that this shows that (ontological) materialism is
false. However, this is a completely unwarranted inference (see
Chapter 7).

3)   Several attempts have been made in the literature on Kant
to clarify his notion of things in themselves. We can distinguish between three different interpretative theories (using my
own terminology): 1) the two-world theory, 2) the double-aspect
theory, and 3) the no-thing theory. According to 1) noumena are
things which are entities of a different, non-sensible kind than
phenomena. The double-aspect theory refers to noumena and phenomena as standing for two different ways of describing or approaching things in the world. Pippin (1982) provides an up-to-
date discussion and evaluation of these views. Pippin's own view
is what I call the no-thing theory. The idea here is to make the
Kantian distinction between constitutive and regulative use of
language and take only constitutive use to be ontological use.
Noumena would fall within the realm of the regulative use of
language, and there would just amount to "regulative ideas" with
nothing in the world really corresponding to them.
     The fourth view, which I adopt, is to give up some of the
basic Kantian assumptions (such as his restriction of human
intuitions to what is sensible) and in a sense replace noumena
by the objects of best-explaining scientific theories.

4)   I have started with the thesis, justified in Chapter 4, that realism is right, and then claimed that adequate realism must be internal or epistemic. This line of thought can be reversed. Thus Rosenberg (1980) shows that a supporter of the internal perspective must be a realist if he rejects the Myth of the Given (MG_e).

Putnam (1982b) on the other hand argues, along Wittgenstein's criticism of solipsism, in the following fashion. He uses the expression 'cultural relativism' which denotes a kind of internal perspective, and demonstrates essentially the following. If (1) my world is conceptually constituted on the basis of my culture, (2) your world is conceptually constituted on the basis of your culture, and if (3) you and I are related symmetrically, then my world and your world will coincide, i.e, realism is correct. The premises can be regarded as true and hence the conclusion as well.

5)   Hellman (1983) discusses several interesting senses of mind-independence. I cannot here comment on his views, which differ somewhat from mine. Let me only point out that he explicates mind-independence most centrally by means of a principle of anti-determination (of the physical by the mental). Put verbally, this principle says that "there is at least a pair of models (meeting whatever standardness requirements you like) of the totality of scientific laws (known or not but formulable in our overall scientific language) which agree on the truth-values of all sentences in phenomenal vocabulary but which differ in the truth-values of some sentences in physical vocabulary" (Hellman, 1983, p. 236). While I agree with this basically I cannot accept Hellman's formalization of it as it a) uses material implication, b) is based on a very strong and non-asymmetric (!) notion of determination (see Hellman and Thompson, 1975).

6)   Sellars, 1963a, 1968, 1979, and 1983, as well as Rosenberg, 1974, contain a fuller presentation and I don't claim to be completely faithful to Sellars' view in all respects. I shall below somewhat modify Sellars' way of speaking and speak of picturing as a causal (or causally grounded) relation between sentences and "the parts of the world they are about". However, this can without further ado be rephrased into Sellars' account according to which natural-linguistic objects picture non-lin-

guistic objects, accepting the latter account as philosophically basic.

7) It is accordingly worth emphasizing that although the notion of picturing is causally grounded on ought-to-be rules it is not required that every correct natural-linguistic picturing-episode be somehow directly causally produced by what it pictures.

8) Sober (1982) argues that Tarski's theory of truth is objectionable in the sense of being too holistic. In Tarski's account, a general sentence is true if and only if it is satisfied by all sequences (or any sequence) of objects. Therefore all true general sentences can be said to be satisfied by the same thing: they are satisfied by all sequences of objects, viz. the whole world. It follows that all true general sentences (cf. "All ravens are black" and "All copper samples expand when heated") bear the same correspondence relation (viz. Tarskian satisfaction) to the world. But a picturing theorist can answer this charge of holism. For surely different picturing relations (and thus specifically different relevant factual world-language relations) are involved in the detailed definition of truth in the case of such different sentences as "All ravens are black" and "All samples of copper expand when heated". Thus even if a picturing theorist holistically defines the truth of general sentences analogously with Tarski (but uses picturing rather than satisfaction in the case of noncompound singular sentences and formulas) he is able to specify a local language-world relation of the kind Sober requires.

CHAPTER 7

1) In saying this we abstract from human limitations - genetic and other - and speak of reason and the scientific method in this generalized sense.

2) Cornman (1975), p. 266, restricts the properties moderate scientific realism is about to "a posteriori" ones. By an a posteriori property (or relationship) he means "one that it is reasonable to claim entities have or lack only if there is some experiential evidence or theoretical scientific reason sufficient to justify the claim". But I do not think there are any

other   real descriptive properties and relations than a   posteriori ones.   (Here I speak of properties and relations in a sense
which   does not commit me to Platonism or anything   resembling.)
But Cornman's own example of 'being owned by someone' - meant as
a counterexample to scientia mensura - certainly does not qualify as an a priori counterexample to that thesis,   for its applicability requires at least experiential evidence.   Thus it seems
that Cornman's qualification is not needed.

     Stich (1983),   p. 222, also argues along somewhat the above
Cornmanian lines: "Consider, for example, such terms as 'favored
by Elizabeth I', 'slept in by George Washington', or 'looks like
Winston   Churchill'.   Surely   none of these terms occurs in   any
currently received scientific theory. Nor is it likely that they
will find a place in the scientific canon of the future.   But it
would be simply perverse to deny,   on these grounds,   that there
are   any   beds   slept in by George Washington   or   any   men   (or
statues)   that   look   like Winston Churchill."   I   agree   to   the
extent that with respect to the manifest image such entities can
be   taken   to   exist.   But such "perversity" gives   no   argument
against   the   scientia mensura claim and the   claim   that   there
really   are no such entities.   For what best-explaining   science
does   is just to redescribe beds,   etc.   in a more accurate   and
truthlike way (this is what 'really' amounts to),   and note that
there need of course be no single term or predicate in the best-
explaining science to represent a bed, for example.

3)   Hence   we can reply to Rorty (1979b,   Chapter VI)   when   he
criticizes   Sellars'   notion of picturing.   According to   Rorty
there are no "good-making" states of affairs for such   sentences
as   "X is good",   and therefore,   presumably,   no "truth-making"
states of affairs. But this criticism overlooks the asymmetry we
have discussed: descriptive talk is ontological, normative discourse is not. Given this, Rorty's argument is not sound.

4)   Recall here the sensa postulated by Sellars   (1971).   They
are   entities in the scientific image postulated   in   accordance
with the scientia mensura -thesis.   In his terminology, they are
physical$_1$ (part of the causal order,   something required for the
best   explanation   of non-living or living matter) even   if   not
physical$_2$ (necessary and sufficient for the best explanation of
non-living matter).

5)   Popper's well-known ontological distinction between the
three "worlds" can be mentioned as a modern example of a way of
dealing with ontological problems and the problem of universals
which leads to muddles. Let it be specifically noted that Popper's Platonistic World Three (the world of propositions, theories, mathematics, works of art, etc.) reminds of Hegel's
Objective Spirit, and that it can be subjected to criticism
analogous to that applicable to Hegel's objective idealism (see
Bunge's 1980, Chapter 8, criticism).

      Sellarsian semantics can be used to justify the claim that
World Three does not exist as an ontological category. Thus,
e.g., the (mathematical) notion of set can in principle be analyzed by help of the relation symbol 'ε' and by referring to the
rules which govern its use. No ontological commitments to sets
ensue from this.

6)   The distinction between psychological and materialistic (or
physicalistic) predicates is not, however, essentially different
from, say, the distinction between biological and physical (or
chemical) predicates. (Recall what I said previously of the
metalinguistic nature intentionality, the feature which is traditionally central to some psychological predicates.)

7)   See Stich, 1983, for a comprehensive recent discussion and
airing of these issues.

                          CHAPTER 8

1)   In Tuomela (1977) a partial analysis of intending is given.
It may be summarized (with two light changes) as follows for so-
called complex intending (cf. Tuomela, 1977, p. 133, and Tuomela, 1984, p. 82; also cf. Audi, 1973):

      An agent A **intends to perform** X **by performing** Y **only if**
      (1) A believes that he, at least with some nonnegligible
          probability, can perform X by performing Y (or at
          least can learn so to perform X by performing Y);
      (2) A wants (and has not temporarily forgotten that he
          wants) to perform X by performing Y;
      (3) A has no equally strong or stronger incompatible want
          (or set of incompatible wants whose combined strength

is at least as great), or, if A does have such a want
or set of wants, he has temporarily forgotten that he
wants the object(s) in question, or does not believe
he wants the object(s), or has temporarily forgotten
his belief that he cannot both realize the object(s)
and perform X by performing Y.

2)   Aulin (1982) has investigated actions in a somewhat similar
systems-theoretic framework as ours. However, we shall not here
go into detailed comparisons but invite the reader to study
Aulin's development for himself.

In developing the technical machinery below suggestions and
criticisms due to Antti Hautamäki were helpful.

3) It is worth pointing out that both the jointness function j_A
and its counterpart J_A could alternatively be defined to account
for more complex cases. We recall that j_A was above taken to
represent joint action. That notion could be taken, for instance, in the broad sense (5.8) of Tuomela (1984) or in some
stricter sense such as (PCS) above. Whatever exact sense is
given to j_A we could anyway define J_A differently from what was
done here. Thus J_A could be allowed to take into account other
features in the world than merely those that $f_{U,A}$ as constrained
by j_A gives. In that way the states $x_A(t)$ could become more
dependent on the environment of the system (e.g., on the social
structures in which it is embedded). Here we cannot, however,
technically investigate these possibilities.

                            CHAPTER 9

1)   The logico-conceptual validity of schema (2), relative to
some rationality assumptions, can be seen by using the following
simple symbolic representation:

(2*)(i*)   I(Y)
    (ii*)   B(U → Y)
    (iii*) I(U/A & K)
    (iv*)   B(A)
    (v*)   B(K)
    ─────────────────
    (vi*)   I(U)

Here I and B refer to a we-intention and a we-belief, respectively, Y to the acquisition of a exhaustively explaining theory, U to the act of replacing a theory by a corrective explanatory theory, and A and K to the contents of the beliefs in the premisses (iv) and (v).

The notion of a conditional intention to perform an action includes that if the agent believes that its conditions are fulfilled he - if he is rational - also intends to perform the act U. Thus (iii*), (iv*) and (v*) logically imply the intention (vi*) of a rational agent. If the agent then in fact does perform U, this act (subjectively, on the level of belief) at least contributes (symbolized by →) towards the realization of the intention in premiss (1).

We can also consider a strong interpretation for the expression 'at least contributes towards' in premiss (ii), according to which the performance of U is necessary for the performance of Y. Then, for a rational agent, (i*) and (ii*) also imply logically the intention (vi*) and hence (vi*) would obtain, so to say, a conclusive justification both from the top downwards and from the bottom upwards.

2) The translation function **tr** can be defined in terms of **general logic** (cf. Feferman, 1974). To sketch that we let S be our predecessor theory in a language L_S and T a successor theory in L_T. The languages L_S and L_T may differ in their purported ontology. Nevertheless the function **tr** will in a suitable model-theoretic sense translate L_S into L_T. Let us see how.

A general logic, say **L**, can be regarded as a quadruple L = $\langle Typ_L, Sent_L, Str_L, \vDash_L \rangle$. Here Typ_L is a class of similarity types for **L**. Roughly speaking, a similarity type, s, consists of a set of symbols standing for or representing "sorts" (kinds of domains), relations, operations, and individual constants. (**L** is typically a many-sorted logic.) $Sent_L$ and Str_L are operations on Typ_L such that $Sent_L(s)$ and $Str_L(s)$ stand for the classes of sentences and structures (possible models, say **M**, of type s∈ Typ_L), respectively, of **L**. (Notice that we may conceptually start with a class of structures and then associate a logic with it: the structures that we are dealing with might be merely conceptual, lacking ontic import, as is the case when they deal with holistic social entities.) Finally, $\vDash_L$ is an operation on Typ_L such that $\vDash_L(s)$ stands for the satisfaction relation of **L** related to **s**. (We will omit **s** and write **M**$\vDash_L$ for $\langle M, \phi \rangle \in \vDash_L(s)$.) We

shall   not   discuss **L** in more detail here except   for   requiring
that it be regular in the sense of Feferman (1974) and   pointing
out that ordinary first-order logic and its usual extensions are
included among such general logics.

Let us now consider our case of translating the predecessor
theory   S into the successor theory **T**. We assume that L_S and L_T
can   be   discussed from the point of view of a general   logic   **L**
which   admits   both models of S and of T as its   structures   and
which, of course, admits S and T as its sentences. **L** may here be
regarded as a logic which has among its class of sentences   both
L_S and L_T. We may indeed make $L_S = \text{Sent}_L(s)$ and $L_T = \text{Sent}_L(t)$, with
the   similarity types **s**,   representing the predecessor's domain,
and **t**,   representing the successor's domain. Now we let M_t be a
model of L_T,   with **t** as its similarity type. Correspondingly, we
let M_s be a model of L_S,   with the similarity type **s**. In general
M_t and M_s may be completely different.

Let us now assume that L_S and L_T are comparable not only in
the weak sense of being based on the same general logic but that
they   are extensionally comparable also in the following   sense.
The models of L_S and L_T can be correlated by a relation,   say **R**,
such that the domain of R belongs to the models of L_T while   its
range   is the class of models L_S. Thus every model M_s of L_S is
assumed   to have a counterpart model M_t of L_T. Technically,   we
let $R \subseteq \text{Str}_L(t) \times \text{Str}_L(s)$,   and   let $\text{Str}_L(s)$ be just the   class   of
possible   predecessor-models   and $\text{Str}_L(t)$ to be   the   class   of
possible   successor-models.   (How to define **R** is of course   very
central but as the matter is strongly context-dependent we shall
not discuss it here.)

The   following   definition   of   translation   connects   the
counterpart   relation **R** with a language and with claims relating
the predecessor's and the successor's realm. We propose:

(**TR**)   Let   **L** be the general logic and let s, $t \in \text{Typ}_L$ and L_S =
    $\text{Sent}_L(s)$, $L_T = \text{Sent}_L(t)$, $R \subseteq \text{Str}_L(t) \times \text{Str}_L(s)$ such that
    R is defined onto $\text{Str}_L(s)$. Then **L** effects a **translation**
    of L_S into L_T relative to **R** if and only if for all $\phi \in L_S$
    there is a $\psi \in L_T$ such that for all **M, N**
    $R(M,N) \Rightarrow [M \vDash_L \psi \Longleftrightarrow N \vDash_L \phi]$.

The notion of translation defined by (**TR**) is a weak one.   First,
it deals only with the preservation of truth but it says nothing
about   meanings.   Secondly,   it   does not in any way define   the

primitive concepts of L_S in terms of those of L_T nor does it require each atomic formula of L_S to have an R-correlate in L_T. Strengthening (**TR**) is somewhat complicated and I shall not discuss it here. Let me, however, emphasize that if we are to get a translation **function** on the basis of (**TR**) then for each $\phi \in L_S$ there must be a **unique** sentence $\psi \in L_T$.

Our present (**TR**) gives a general solution to the translation problem for statements sharing an underlying general logic, for it claims that for each general sentence ϕ (e.g., the whole theory S) there exists an R-counterpart ψ (e.g., T) and it puts no conditions on the vocabularies in question (any more than **R** does). This means that, if (**TR**) is satisfied, we can translate each sentence (e.g., theory) of L_S into L_T, and we may then go ahead and compare theories stated in L_T as to their mutual distances, for instance. When **L** is a first-order logic the measures of Tuomela (1978b) become applicable.

Note that whenever S and T are both assumed "given" prior to, and independently of, the translation, our (**TR**) need not translate (or reduce, we may even say) S into T even if it does translate L_S into L_T. This is because **R** has been defined only generally for models of L_S and L_T, and it need not correlate each model of S with a model of T in cases when T is not a translation of S by definition. If that is regarded as a problem the remedy is obvious: require that **R** always correlates models of S with models of T only (such that **R** is a relation onto the class of models of S). I shall not explicitly spell this out here. As shown by Feferman (1974), the conditions of (**TR**) can indeed be satisfied, and translation effected, in a great variety of cases.

In general there are no a priori guarantees that there be only one such translation function. Thus our account accepts the possibility of there being indeterminacy of translation in something like the Quinean sense. On the other hand, if our sketchy remarks about the evolutionary character of the comparative notion of picturing (viz., "more adequate picturing") in Chapter 6 are on the right track there will be no a priori proof that interlinguistic picturing and translation are indeterminate. Rather it can be expected on a posteriori grounds that these notions, which in a way are the two sides of the same coin (these sides of the coin being the intercomparison of conceptual schemes, and the naturalistic notion of comparative picturing on the other hand), will be, or are construable as more or less determinate, at least if there will be a unique limit science.

It should be noted here that our definitions of the notions
of better and best explanation to be given below have been
stated without relativization to translation, viz. as if translation were determinate or as if the best translation were at
hand. In a finer analysis the possibility of correcting predecessor theories with successor theories in several different
ways should be made more explicit by relativization to tr.

3) Let me, however, point out that our notion of corrective
explanation is compatible with the indeterminacy of translation
function tr (if indeed it is indeterminate). By indeterminacy I
mean the possibility of there being several translation functions preserving corrective explanation in the defined sense.
Note that here T_{i+1} will explain the approximate truth (and
success) of T_i rather than its truth (or whatever true content
it has). The true statements of T_i are not required to be
strictly preserved by translation. Thus, contra e.g. Sellars
(1968), we allow for loss of truth here. This is a possible
position, as we do not think that science must proceed towards a
Peircean limit science (or anything resembling). Our account
accordingly escapes the criticisms of Burian (1979).

4) Note that I here accept a metalinguistic approach to lawlikeness (theoryhood, nomicity). Thus the object language contains no intensional operators (such as 'nomically implies')
expressing lawlikeness. Instead we classify the generalizations
of the object language in our metalanguage by means of predicates such as 'is lawlike', 'is a theory', and so on (e.g. 'T is
a theory', 'T^* is not a theory').

Furthermore, if T is a theory it must be pragmatically adequate in the sense of its holders' (the members' of the scientific community A) being able, at least under favorable conditions, to perform appropriate metalinguistic inferences involving the predicates of the theory (thus to infer, to take an
example, 'a will expand' from 'a is copper & a is heated' in the
relevant theory relating the heating and expanding of copper).

5) Our present view of theories is compatible with the systems-theoretic state-space account of theories (cf. Chapter 8).
This can be seen by following Carnap's later work related to
inductive logic. For we may suitably collect the monadic predi-

cates of our language $L(\lambda U\{R\})$ into families. Let $P_{i,j}$, $i = 1,\ldots,n$, $j=1,\ldots,k(i)$, be our predicates. Then each family $F_i = \{P_{i,1},\ldots, P_{i,k(i)}\}$ is such that the predicates in it are mutually exclusive and jointly exhaustive (normally for conceptual reasons). Now the n families F_i, $i=1,\ldots,n$, generate an n-dimensional conceptual space, S, where the elements are the so-called Q-predicates defined by the conjunctions $P_{1j(1)}$ & $P_{2j(2)}$ &...& $P_{nj(n)}$, where $1 \leq j(1) \leq k_1$, $\ldots$; $1 \leq j(n) \leq k(n)$.

Given this, we may formulate, e.g., laws of succession by relativizing these Q-predicates to time and claim that a certain Q-predicate exemplified at time t will be followed by an exemplification of a certain other Q-predicate at t+1. What we then have here is a finite version of the state-space approach. Were we to let each F_i contain an uncountably infinite number of predicates, indeed the real axis, we would arrive at the standard state-space approach. (See Niiniluoto, 1985, for a relevant discussion.)

6) Because the examination of the distances between sentences are technical and rather demanding, I cannot here go into the details. Nevertheless I want to allude to a central idea that I have employed earlier. When the sentences S and S' are presented in their distributive normal forms we can see that each one of them generally claims that certain kinds of individuals exist and others do not. About some kinds they refrain from saying anything. If now CT_S and $CT_{S'}$ refer to the kinds allegedly exemplified in the theories or sentences S and S', and CT_S^* and $CT_{S'}^*$ to the corresponding excluded kinds, we reason as follows. The distance between S and S' is on one hand directly related to the size of the symmetric difference of the sets CT_S and $CT_{S'}$, and on the other hand to the size of the symmetrical difference between CT_S^* and $CT_{S'}^*$. Technically the measure d is obtained from these by using suitable normalizations and free parameters (see Tuomela, 1978b).

7) As shown by Hintikka and Niiniluoto (1976) these probability measures can be axiomatized, as far as their formal, non-epistemic features are concerned, as follows. We assume that P_A is a real-valued function defined for pairs of sentences (h,e) of a language with k primitive predicates and N individual constants (such that N may be infinite). P_A is then assumed to satisfy the following axioms:

(A1)   Probability axioms (in the standard sense).

(A2)   Finite regularity:
For singular sentences h and e, $P_A(h/e) = 1$ only if e logically implies h.

(A3)   Symmetry with respect to individuals:
The value of $P_A(h/e)$ is invariant with respect to permutation of individual constants.

(A4)   Symmetry with respect to predicates:
The value of $P_A(h/e)$ is invariant with respect to permutation of the Q-predicates.

(A5)   c-principle:
There is a function f such that $P_A(Q_i(a(n+1))/e(c,n)) = f(n_i,n,c)$, where $e(c,n)$ denotes a sample of size n exemplifying c kinds of individuals and where n_i denotes the number of individuals satisfying Q_i.

The measures P_A satisfying the above axioms generate the so-called K-dimensional system of inductive logic. Excluding degenerate cases some general sentences of the language in question can then be shown to receive non-zero probabilities even in infinite universes. (For how to handle the somewhat problematic symmetry axiom (A4) in applications, see the discussion in Niiniluoto and Tuomela, 1973.)

CHAPTER 10

1)   One well known example illustrating the self-correctiveness of the method of science is provided by de Finetti's limit theorem, which we shall below state in the formulation used by Savage (1954), pp. 46 - 50 (also cf. Niiniluoto's, 1980, discussion). Assume thus that $e_1, e_2, \ldots$, are the successive independent results of a random experiment. Let h be a hypothesis with nonzero a priori probability, viz. $P(h) > 0$, such that the likelihood of every e_n under h differs from that under $\sim h$, viz. $P(e_n/h) \neq P(e_n/\sim h)$. Then the following holds, on the basis of probability calculus:

(*) $$P\left(\frac{P(h/e_1 \& \ldots \& e_n)}{P(\sim h/e_1 \& \ldots \& e_n)} \to \infty/h\right) = 1, \text{ when } n \to \infty$$

Defining $h^* = h$, if h is true, and $h^* = \sim h$, if h is false, (*)
entails, if $P(h^*) > 0$,

(**) $P(P(h^*/e_1 \& \ldots \& e_n) \approx 1) \to 1$, when $n \to \infty$

What (**) perspicuously amounts to saying is that
(a) if the true hypothesis is not a priori excluded, viz. if
$P(h^*) > 0$; and if
(b) relevant empirical information (such that $P(e_n/h) \neq P(e_n/\sim h)$)
is gathered to a sufficient extent,
then the weight of such information, assumed to be "exchange-
able" or symmetric, will eventually be overwhelming - it will
reveal the truth (or falsity, as the case may be) of the hypothesis h with probability approaching 1. In saying this I have
interpreted P as a degree of knowledge and assumed probabilistically certain knowledge to entail truth. (The above result is
philosophically interesting - as far as it goes. I shall not
here go into an evaluation of its applicability to actually
used scientific methodology and real scientific research.)

2) A classical and funny description of pseudoscience is given
in Swift's **Gulliver's Travels**, section III, Chapter 5.

3) For instance, Schmidt's mentioned studies of psychokinesis,
which are based on the use of a random generator, may serve as
examples of prescientific research, although, as I think, they
have been criticized with good reason. (Cf. Alcock, 1981, pp.
168 - 179, and Hansel's, Hyman's, and Schmidt's discussion in
The Skeptical Inquirer, vol. 3, 1981.)

4) I shall here just refer, e.g., to the issues of **The Skeptical Inquirer** which have come out during the last few years; in
those issues almost all of the pseudodoctrines, mentioned above,
have been subjected to criticisms; cf. also e.g. Gardner (1957),
(1982), Abell and Singer (1981), Frazier (1981), Grim (1982),
Randi (1982). See Priolean et al. (1983) for comprehensive evidence and defense of the claim that the benefits of psychoterapy
do not statistically exceed those obtained from placebo treat-
ments in the case of neurotic outpatients.

5) Such a scientific attitude is not a version of scientism,
which is in the first place an ideology, an axiological doc-

trine. We have accepted (in Chapter 7) the Kantian distinction
questio facti - quaestio juris, and maintained that the scientific attitude now under consideration is only connected to
ontological matters. Consequently its objects are **questiones
facti**. Value problems are not solved by science but by **us**, by
means of forming we-intentions (and other we-attitudes). This
challenge concerns all people.

Abell, G. and Singer, B. (eds.): 1981, **Science and the Paranormal**, Junction Books, London.

Alcock, J.: 1981, **Parapsychology: Science or Magic?**, Pergamon Press, Oxford.

Aspect, A., Dalibard, J., and Roger, G.: 1982, 'Experimental Test of Bell's Inequalities Using Time-Varying Analyzers', **Phys. Rev. Lett., 49**, 1804 - 1807.

Audi, R.: 1973, 'Intending', **The Journal of Philosophy, 70**, 387 - 403.

Aulin, J.: 1982, **The Cybernetic Laws of Social Progress**, Pergamon Press, Oxford.

de Beauregard, O.C.: 1983, 'Lorentz and CPT Invariance and the EPR Correlations', **Abstracts of the 7th International Congress of Logic, Methodology, and Philosophy of Science, Vol. 4** (1983), 28 - 31.

Bell, J.S.: 1964, 'On the Einstein-Podolsky-Rosen Paradox', **Physics, 1**, 195 - 200.

Bell, J.S.: 1971, 'Introduction to the Hidden-variable Question' in d'Espagnat, B. (ed.), **Foundations of Quantum Mechanics**, Academic Press, New York, 171 - 181.

Boyd, R.: 1981, 'Scientific Realism and Naturalistic Epistemology', in Asquith, P. and Giere, R. (eds.), **PSA 1980, Vol. II**, Philosophy of Science Association, East Lansing, 613 - 662.

Boyd, R.: 1983, 'On the Current Status of the Issue of Scientific Realism', **Erkenntnis, 19**, 45 - 90.

Bradie, M.: 1979, 'Pragmatism and Internal Realism', **Analysis, 39**, 4 - 10.

Braude, S.: 1980, **ESP and Psychokinesis**, Temple University Press, Philadelphia.

Bunge, M.: 1967, **Foundations of Physics**, Springer-Verlag, Berlin, Heidelberg, and New York.

Bunge, M.: 1979, **Treatise on Basic Philosophy, Vol. 4: A World of Systems**, Reidel, Dordrecht.

Bunge, M.: 1980, **The Mind-Body Problem**, Pergamon Press, Oxford.

Bunge, M.: 1981, **Scientific Materialism**, Reidel, Dordrecht.

Bunge, M.: 1983, **Treatise on Basic Philosophy, Vol. 6: Understanding the World**, Reidel, Dordrecht/Boston/Lancaster.

Burian, R.: 1979, 'Sellarsian Realism and Conceptual Change in Science', in Bieri, P., Horstmann, R-P., and Krüger, L. (eds.), **Transcendental Arguments and Science**, Reidel, Dordrecht, 197 - 225.

Cartwright, N.: 1983, **How the Laws of Physics Lie**, Clarendon Press, Oxford University Press, New York, Oxford.

Castaneda, H-N.: 1975, **Thinking and Doing**, Reidel, Dordrecht.

Churchland, P.M., Hooker, C.A. (eds.): 1985, **Images of Science, Scientific Realism vs. Constructive Empiricism**, University of Chicago Press, forthcoming.

Clauser, J.F. and Shimony, A.: 1978, 'Bell's Theorem: Experimental Tests and Implications', **Rep. Prog. Phys., 41**, 1881 - 1927.

Cornman, J.: 1971, **Materialism and Sensations**, Yale University Press, New Haven.

Cornman, J.: 1972, 'Materialism and Some Myths about Some Givens', **The Monist, 56**, 215 - 233.

Cornman, J.: 1975, **Perception, Common Sense, and Science**, Yale University Press, New Haven.

Delaney, C., Gutting, G., Solomon, W., and Loux, M. (eds.): 1977, **The Synoptic Vision: Essays on the Philosophy of Wilfrid Sellars**, University of Notre Dame Press, Notre Dame, In.

Eberhard, P.H.: 1977, 'Bell's Theorem without Hidden Variables', **Nuovo Cimento, 38B**, 75 - 80.

Einstein, A., Podolsky, B, and Rosen, N.: 1935, 'Can Quantum-Mechanical Description of Physical Reality Be Considered Complete?' **Phys. Rev., 47**, 777 - 780.

d'Espagnat, B.: 1979, 'The Quantum Theory and Reality', **Scientific American, 241**, No 5, 128 - 140.

Eysenck, H. and Sargent, C.: 1982, **Explaining the Unexplained**, Multimedia Publications, Willemstad.

Feferman, S.: 1974, 'Two notes on Abstract Model Theory: I. Properties Invariant on the Range of Definable Relations Between Structures', **Fundamenta Mathematicae, 82**, 153 - 165.

Feigl, H.: 1967, **The 'Mental' and the 'Physical'**, University of Minnesota Press, Minneapolis.

Field, H.: 1980, **Science without Numbers. A Defence of Nominalism**, Blackwell, Oxford.

Fodor, J.: 1975, **Language of Thought**, Harvard University Press, New York.

Fodor, J.: 1984, 'Psychosemantics or: Where do Truth Conditions Come From?', forthcoming.

van Fraassen, B.: 1980, **The Scientific Image**, Clarendon Press, Oxford.

Frazer, J.: 1922, **The Golden Bough**, Macmillan, London.

Frazier, K. (ed.): 1981, **Paranormal Borderlands of Science**, Prometheus Books, Buffalo, N.Y.

Gardner, M.: 1957, **Fads and Fallacies in the Name of Science**, Dover, New York.

Gardner, M.: 1982, **Science: Good, Bad, Bogus**, Prometheus Books, Buffalo.

Grattan-Guinness, I. (ed.): 1982, **Psychical Research**, The Aquarian Press, Wellingborough.

Greenwell, J.: 1980, 'Academia and the Occult: An Experience at Arizona', **The Skeptical Inquirer**, 5, 39 - 45.

Grim, P. (ed.): 1982, **Philosophy of Science and the Occult**, State University of New York Press, Albany, N.Y.

Gärdenfors, P.: 1984, 'The Dynamics of Belief as a Basis of Logic', **British Journal for the Philosophy of Science**, 35, 1 - 10.

Habermas, J.: 1979, **Communication and the Evolution of Society**, Beacon Press, Boston.

Hacking, I.: 1983, **Representing and Intervening, Introductory Topics in the Philosophy of Natural Science**, Cambridge University Press, Cambridge.

Hansel, C.: 1980, **ESP and Parapsychology**, Prometheus Books, Buffalo.

Hellman, G.: 1982, 'Stochastic Einstein-locality and the Bell Theorems', **Synthese**, 53, 461 - 504.

Hellman, G.: 1983, 'Realist Principles', **Philosophy of Science**, 50, 227 - 249.

Hellman, G. and Thompson, F.W.: 1975, 'Physicalism: Ontology, Determination, and Reduction', **The Journal of Philosophy**, 72, 551 - 564.

Hempel, C.G.: 1965, **Aspects of Scientific Explanation**, The Free Press, New York.

Hempel, C.G.: 1966, **Philosophy of Natural Science**, Englewood Cliffs, N. J., Prentice-Hall.

Hintikka, J.: 1970, 'Surface Information and Depth Information' in Hintikka, J. and Suppes, P. (eds.), **Information and Inference**, Reidel, Dordrecht, 263 - 297.

Hintikka, J.: 1973a, **Logic, Language Games, and Information: Kantian Themes in the Philosophy of Logic**, Oxford University Press, Oxford.

Hintikka, J.: 1973b, 'On the Different Ingredients of an Empirical Theory', in Suppes, P. et. al. (eds.), **Logic, Methodology, and the Philosophy of Science**, **IV**, North-Holland Publishing Company, Amsterdam, 313 - 322.

Hintikka, J.: 1984, 'Das Paradox transzendentaler Erkenntnis', in Vossenkuhl and Schaper (eds.), **Bedingungen der Möglichkeit**, Klett-Cotta Verlag.

Hintikka, J. and Tuomela, R.: 1970, 'Towards a General Theory of Auxiliary Concepts and Definability in First-Order Theories' in Hintikka, J. and Suppes, P. (eds.), **Information and Inference**, Reidel, Dordrecht, 298 - 330.

Hintikka, J., and Niiniluoto, I.: 1976, 'An Axiomatic Foundation for the Logic of Inductive Generalization', in Przelecki, M., Szaniawski, K., and Wojcicki, R. (eds.), Proceedings of the Conference for **Formal Methods in the Methodology of the Empirical Sciences**, Warsaw, June 17 - 21, 1974, Reidel, Dordrecht, 1976, 57 - 81.

Hooker, C.A.: 1985, 'Surface Dazzle, Ghostly Depths: An Exposition and Critical Evaluation of van Fraassen's Vindication of Empiricism Against Realism' in Churchland, P. M. and Hooker C. A. (eds.).

Husserl, E.: 1913, **Ideen zu einer reinen Phänomenologie und phänomenologischen Philosophie I**, Niemayer, Halle.

Hyman, R.: 1982, 'Does the Ganzfeld Experiment Answer the Critic's Objections?', in **Program and Presented Papers**, Vol. I of the Centenary Jubilee Conference of the Society for Psychical Research, 1882 - 1982, Tues. 17 Aug., 1982, 21 pp.

Jammer, M.: 1974, **The Philosophy of Quantum Mechanics**, Wiley, New York.

Johnson, M.: 1980, **Parapsykologi**, Zindermans, Uddevalla.

Kant, I.: 1781/1787, **Kritik der reinen Vernunft**, Hartknoch, Riga.

Kitcher, P.: 1980a, 'A Priori Knowledge', **Philosophical Review**, **89**, 3 - 23.

Kitcher, P.: 1980b, 'Apriority and Necessity', **Australian Journal of Philosophy**, **58**, 89 - 101.

Koethe, J.: 1979, 'Putnam's Argument against Realism', **The Philosophical Review, 88**, 92 - 99.

Kripke, S.: 1972, 'Naming and Necessity', in Davidson, D. and Harman, G. (eds.), **Semantics of Natural Language**, Reidel, Dordrecht, 253 - 355, 763 - 769.

Laudan, L.: 1973, 'Peirce and the Trivialization of the Self-Correcting Thesis', in Giere, R., and Westfall, R. (eds.), **Foundations of Scientific Method: The 19th Century**, Indiana University Press, Bloomington, 275 - 306.

Laudan, L.: 1981, 'A Refutation of Convergent Realism', in Jensen, U.J. and Harré, R. (eds.), **The Philosophy of Evolution**, Cambridge University Press, Cambridge, 232 - 268.

Laudan, L.: 1984, **Science and Values**, University of California Press, Berkeley, Los Angeles, London.

Lewis, D.: 1984, 'Putnam's Paradox', **Australian Journal of Philosophy, 62**, 221 - 236.

Loux, M.: 1977, 'Ontology', in Delaney, C., Gutting, G., Solomon, W., and Loux, M. (eds.), 43 - 72.

Loux, M.: 1978, 'Rules, Roles, and Ontological Commitment: An Examination of Sellars' Analysis of Abstract Reference', in Pitt J. (ed.), 229 - 256.

Mackie, J.: 1977, **Ethics**, Penguin Books, Middlesex.

Mackie, J.: 1982, **The Miracle of Theism**, Clarendon Press, Oxford.

Marks, D. and Kammann, R.: 1980, **The Psychology of the Psychic**, Prometheus Books, Buffalo.

Marras, A.: 1978, 'Rules, Meaning and Behavior: Reflections of Sellars' Philosophy of Language', in Pitt, J. (ed.), 163 - 187.

Matson, I.: 1966, 'Why Isn't the Mind-Body Problem Ancient', in Feyerabend, P., and Maxwell, G. (eds.), **Mind, Matter, and Method**, Minnesota University Press, Minneapolis, 92 - 102.

Mittelstaedt, P. and Stachow, E.W.: 1983, 'Analysis of the Einstein-Podolsky-Rosen Experiment by Relativistic Quantum Logic' **Int. J. Theor. Phys., 22**, 517 - 540.

Musgrave, A.: 1984, 'Constructive Empiricism and the Theory-Observation Dichotomy', **Abstracts of the 7th International Congress of Logic, Methodology and Philosophy of Science, Vol. 6** (1983), p. 149.

Musgrave, A.: 1985, 'Realism Versus Constructive Empiricism', in Churchland, P. M. and Hooker, C. A. (eds.).

Niiniluoto, I.: 1977, 'On the Truthlikeness of Generalizations', in Butts, R., and Hintikka, J. (eds.), **Basic Problems in Methodology and Linguistics**, Reidel, Dordrecht, 121 - 147.

Niiniluoto, I.: 1978, 'Truthlikeness: Comments on Recent Discussion', **Synthese**, **38**, 281 - 330.

Niiniluoto, I.: 1980, 'Scientific Progress', **Synthese**, **45**, 427 - 462.

Niiniluoto, I.: 1985, 'Theories, Approximations, and Idealizations', in Cohen, L.J. et. al. (ed.), **Logic, Methodology, and Philosophy of Science VII**, North-Holland, Amsterdam, forthcoming.

Niiniluoto, I., and Tuomela, R.: 1973, **Theoretical Concepts and Hypothetico-Inductive Inference**, Reidel, Dordrecht.

Pearce, D., and Rantala, V.: 1982a, 'Realism and Formal Semantics', **Synthese**, **53**, 39 - 53.

Pearce, D., and Rantala, V.: 1982b, 'Realism and Reference: Some Comments on Putnam', **Synthese**, **52**, 439 - 448.

Pippin, R.B.: 1982, **Kant's Theory of Form, An Essay on the Critique of Pure Reason**, Yale University Press, New Haven and London.

Pitt, J. (ed.): 1978, **The Philosophy of Wilfrid Sellars: Queries and Extensions**, Reidel, Dordrecht.

Post, J. F.: 1983, 'Correspondent Truth without Determinate Reference', **Abstracts of the 7th International Congress of Logic, Methodology, and Philosophy of Science, Vol. 2** (1983), 139 - 143.

Priolean, L., Murdock, M., and Brody, N.: 1983, 'An Analysis of Psychoterapy versus Placebo Studies', **The Behavioral and Brain Sciences**, **6**, 275 - 310.

Putnam, H.: 1975, **Mind, Language, and Reality**, Cambridge University Press, Cambridge.

Putnam, H.: 1978, **Meaning and the Moral Sciences**, Routledge and Kegan Paul, London.

Putnam, H.: 1981, **Reason, Truth, and History**, Cambridge University Press, Cambridge.

Putnam, H.: 1982a, 'Why There Isn't a Ready-Made World', **Synthese**, **51**, 141 - 168.

Putnam, H.: 1982b, 'Why Reason Can't Be Naturalized', **Synthese**, **52**, 3 - 23.

Putnam, H.: 1983, 'Vagueness and Alternative Logic', in **Realism and Reason**, Cambridge University Press, Cambridge, Massachusetts, 271 - 286.

Putnam, H.: 1984, 'A Comparison of Something with Something Else', forthcoming in **New Literature History.**

Quine, W.V.: 1969, **Ontological Relativity and Other Essays,** Columbia University Press, New York.

Radner, D. and Radner, M.: 1982, **Science and Unreason,** Wadsworth, Belmont.

Randi, J.: 1982, **Flim-Flam!,** Prometheus Books, Buffalo.

Rescher, N.: 1977, **Methodological Pragmatism,** New York University Press, New York.

Rescher, N.: 1978, **Peirce's Philosophy of Science,** University of Notre Dame Press, Notre Dame.

Rorty, R.: 1970, 'Strawson's Objectivity Argument', **Review of Metaphysics, 24,** 207 - 244.

Rorty, R.: 1979a, 'Epistemological Behaviorism and the De-Transcendentalization of Analytic Philosophy', **Neue Hefte für Philosophy, Heft 14: Zur Zukunft der Transzendentalphilosophie,** 115 - 142.

Rorty, R.: 1979b, **Philosophy and the Mirror of Nature,** Princeton University Press, Princeton.

Rosenberg, J.: 1974, **Linguistic Representation,** Reidel, Dordrecht-Holland/Boston-U.S.A.

Rosenberg, J.: 1980, **One World and Our Knowledge of It,** Reidel, Dordrecht.

Russell, B.: 1917, **Mysticism and Logic,** Doubleday Anchor Book, New York 1957.

Russell, J.B.: 1972, **Witchcraft in the Middle Ages,** Cornell University Press, Ithaca and London.

Ryle, G.: 1948, **The Concept of Mind,** Hutchinson, London.

Savage, L.J.: 1954, **The Foundations of Statistics,** John Wiley and Sons, New York.

Sellars, W.: 1948a, 'Realism and the New Way of Words', **Philosophy and Phenomenological Research, 8,** 601 - 634.

Sellars, W.: 1948b, 'Concepts as Involving Laws and Inconceivable without Them', **Philosophy of Science, 15,** 289 - 315.

Sellars, W.: 1963a, **Science, Perception, and Reality,** Routledge and Kegan Paul, London.

Sellars, W.: 1963b, 'Empiricism and Abstract Entities' in Schilpp, P. (ed.), **The Philosophy of Rudolf Carnap,** Open Court, La Salle, 431 - 468.

Sellars, W.: 1967a, **Philosophical Perspectives,** Charles C. Thomas, Springfield.

Sellars, W.: 1967b, 'Some Remarks on Kant's Theory of Experi-
    ence', **Journal of Philosophy, 64,** 633 - 647. (Also in Sel-
    lars, W.: 1974.)
Sellars, W.: 1968, **Science and Metaphysics,** Routledge and Kegan
    Paul, London.
Sellars, W.: 1971, 'Science, Sense Impressions, and Sense: A
    Reply to Cornman', **The Review of Metaphysics, 23,** 391 - 447.
Sellars, W.: 1972, '... this I or he or it (the thing) which
    thinks ...', **Proceedings of the American Philosophical Asso-
    ciation, 44,** 5 - 31. (Also in Sellars, W.: 1974.)
Sellars, W.: 1974, **Essays in Philosophy and Its History,** Reidel,
    Dordrecht.
Sellars, W.: 1977, 'Is Scientific Realism Tenable?' in Suppe,
    F., and Asquith, P., (eds.), **PSA 1976,** Vol. II, 307 - 334.
Sellars, W.: 1979, **Naturalism and Ontology,** Ridgeview, Reseda.
Sellars, W.: 1980, 'Behaviorism, Language, and Meaning', **Pasific
    Philosophical Quarterly, 61,** 3 - 25.
Sellars, W.: 1981, 'Mental Events', **Philosophical Studies, 39,**
    325 - 345.
Sellars, W.: 1983, 'Towards a Theory of Predication', in Bogen,
    J. (ed.), **How Things Are,** Reidel, Dordrecht.
Simon, H.A.: 1970, 'The Axiomatization of Physical Theories',
    Philosophy of Science, 37, 16 - 26.
Sober, E.: 1982, 'Realism and Independence', **Noûs, 16,** 369 -
    385.
Stich, S.P.: 1983, **From Folk Psychology to Cognitive Science,**
    The MIT Press, Cambridge, Massachusetts, London, England.
Suppes, P. and Zanotti, M.: 1980, 'A New Proof of the Impossi-
    bility of Hidden Variables using the Principles of Exhange-
    ability and Identity af Conditional Distributions', in Sup-
    pes, P. (ed.), **Studies in the Foundations of Quantum Mechan-
    ics,** Philosophy of Science Association, East Lansing, Michi-
    gan, 173 - 191.
Swinburne, R.: 1979, **The Existence of God,** Clarendon Press, Ox-
    ford.
Thomson, M.: 1972, 'Singular Terms and Intuitions in Kant's
    Epistemology', **Review of Metaphysics, 26,** 314 - 343.
Tuomela, R.: 1973, **Theoretical Concepts,** Springer, Wien, and New
    York.
Tuomela, R.: 1976, 'Causes and Deductive Explanation', in Cohen,
    R., Hooker, C., Michalos, A., and Evra, J. (eds.), **PSA 1974,**
    Reidel, Dordrecht, 325 - 360.

Tuomela, R.: 1977, **Human Action and Its Explanation**, Reidel, Dordrecht and Boston.

Tuomela, R.: 1978a, 'Scientific Realism and Perception', **British Journal for the Philosophy of Science, 29**, 87 - 104.

Tuomela, R.: 1978b, 'Theory-Distance and Verisimilitude', **Synthese, 38**, 213 - 246.

Tuomela, R.: 1979, 'Putnam's Realisms', **Theoria, 45**, 114 - 126.

Tuomela, R.: 1980a, 'Analogy and Distance', **Zeitschrift für Allgemeine Wissenschaftstheorie, 11**, 276 - 291.

Tuomela, R.: 1980b, 'Explaining Explaining', **Erkenntnis, 15**, 211 - 243.

Tuomela, R.: 1981, 'Inductive Explanation', **Synthese, 48**, 257 - 294.

Tuomela, R.: 1984, **A Theory of Social Action**, Reidel, Dordrecht, Boston, and Lancaster.

Tuomela, R and Miller, K.: 1985, 'We-intentions and Social Action', forthcoming in **Analyse und Kritik**.

Ullman, S.: 1980, 'Against Direct Perception', **The Behavioral and Brain Sciences, 3**, 373 - 415.

Wittgenstein, L.: 1953, **Philosophical Investigations**, Blackwell, Oxford.

Wolman, B. (ed.): 1977, **Handbook of Parapsychology**, Van Nostrand, New York.

von Wright, G.H.: 1971, **Explanation and Understanding**, Cornell University Press, Ithaca.

Abell, G. 231, 255

Apel, K-O. 5

Alcock, J. 221 - 223, 231 - 233, 255

Aristotle 11, 135, 137 - 138

Aspect, A. 57

Audi, R. 247

Aulin, Y. 162, 165, 248

de Beauregard, O.C. 62

Bell, J. S. 48, 52 - 58, 60 - 63, 241

Bohr, N. 53, 64

Boyd, R. 46, 51 - 52, 217, 240

Bradie, M. 240

Braude, S. 227

Brentano, F. 30

Bunge, M. 13, 19, 87, 134, 138, 153, 215, 217, 219, 227, 247

Burian, R. 119, 252

Carnap, R. 2, 252

Cartwright, N. 44

Castañeda, H-N. 130

Churchland, P.M. 239

Clauser, J.F. 54, 56 - 57, 63

Cornman, J. 25, 27, 65, 69 - 71, 79, 108, 125 - 128, 134, 245 - 246

Craig, W. 66 - 68, 79, 81

Descartes, R. 136, 137

Dewey, J. 2

von Däniken, E. 229

Eberhard, P.H. 57

Eddington, A. 42, 49, 125, 128

Einstein, A. 52 - 53, 60, 63

d'Espagnat, B. 57, 63

Eysenck, H. 231

Feferman, S. 249 - 251

Feigl, H. 135

Feyerabend, P. 14

Field, H. 85

Fodor, J. 116

van Fraassen, B. 41 - 43, 48, 51, 67, 77, 239 - 240

Frazer, J. 221 - 222

Frazier, K. 231, 255

Gadamer, H-G. 5

Gärdenfors, P. 237 - 238

Gardner, M. 231, 255

Gödel, K. 45

Grattan-Guinness, I. 231

Greenwell, J. 226

Grim, P. 255

Habermas, J. 5

Hacking, I. 44

Hansel, C. 231, 255

Hanson, N.R. 41

Hautamäki, A. 248

Hegel, G.W.F. 134, 247

Heidegger, M. 5

Hellman, G. 61 - 62, 110, 244

Hempel, C.G. 182, 241

Hintikka, J. 1, 80 - 81, 87, 89 - 90, 182, 190 - 191, 193 - 194, 197, 236, 253

Hooker, C.A. 239

Hume, D. 6, 14, 27

Husserl, E. 2, 5, 36, 97

Hyman, R. 231 - 232, 255

Jammer, M. 54

Johnson, M. 231

Jung, C.G. 223

Kammann, R. 231 - 232

Kant, I. 1 - 2, 4, 31, 37 - 40, 95, 100 - 103, 107 - 108, 129, 225, 236, 238 - 239, 243

Kitcher, P. 236

Koethe, J. 96

Kripke, S. 136, 236

Kuhn, T. 41, 52, 122, 170, 172

Laudan, L. 201 - 209, 213, 240

Lewis, D. 96

Locke, J. 6, 27

Loux, M. 238

Mackie, J. 130, 224

Marks, D. 231 - 232

Marras, A. 238

Matson, I. 137

Miller, K. 147

Mittelstaedt, P. 59 - 61

Montaque, R. 62

Moore, G.E. 14

Musgrave, A. 239

Niiniluoto, I. 78 - 79, 81, 175, 190 - 191, 195, 197, 199, 203, 253 - 254

Pearce, D. 27, 96

Peirce, C.S. 2, 12, 213

Piaget, J. 34, 121

Pippin, R.B. 243

Plato 17, 135, 137

Podolsky, B. 52

Popper, K. 247

Post, J.F. 100

Priolean, L. 255

Puthoff, H. 231

Putnam, H. 26 - 27, 29 - 30, 40, 46, 95 - 101, 104, 106, 114, 117, 119, 122, 201, 203 - 204, 240, 243 - 244

Quine, W.V. 26, 98

Radner, D. 227

Radner, M. 227

Ramsey, F.P. 66, 68

Randi, J. 231 - 232, 255

Rantala, V. 27, 96

Rescher, N. 213 - 214

Rorty, R. 2, 5, 41, 131, 135 - 138, 239, 246

Rosen, N. 52

Rosenberg, J. 39, 101, 103, 114, 170 - 171, 199, 244

Russell, B. 2, 90

Russell, J.B. 221

Ryle, G. 136

Sargent, C. 231

Sartre, J.P. 131

Savage, L.J. 254

Schmidt, H. 231 - 232, 255

Sellars, W. 2, 5 - 6, 8, 10 - 12, 14 - 17, 20, 22 - 23, 25 - 28, 30 - 32, 34 - 36, 38 - 39, 41 - 44, 50 - 51, 67 - 69, 89, 100 - 103, 116, 118 - 120, 122, 125, 130, 138, 169 - 170, 191 - 192, 238 - 239, 241, 244, 246, 252

Shimony, A. 54, 56 - 57, 63

Shoemaker, S. 136

Simon, H.A. 85

Singer, B. 231, 255

Skinner, B.F. 222

Smart, J.J.C. 132

Sober, E. 245

Stachow, E.W. 59 - 61

Stich, S.P. 237, 246 - 247

Suppes, P. 57

Swift, J. 255

Swinburne, R. 224

Targ, R. 231

Tarski, A. 122, 245

Thompson, F.W. 110, 244

Thomson, M. 238 - 239

Tuomela, R. 17, 20, 26, 34, 48, 51 - 52, 66 - 67, 70 - 71, 73, 78 - 82, 87, 96, 99, 102, 111, 123, 127 - 128, 131, 141, 143 - 152, 158, 161, 170, 172, 174 - 175, 178, 182 - 183, 187 - 188, 190 - 191, 196 - 199, 203, 240 - 241, 247 - 248, 251, 253 - 254

Ullman, S. 116

Wegener, A. 234

Wittgenstein, L. 2, 31, 35, 116, 136, 244

Wolman, B. 231

von Wright, G.H. 142

Zanotti, M. 57

action
   intentional action **144**
   joint social action **145 - 146**
action-consequences **141 - 142**
action-results **142**
analogy theory of thinking 17 - 18
aqua regia -example 47ff., 68ff., 240 - 241
arguments for scientific realism 42ff.
   de **facto** postulation of unobservable entities by science 45 -
   46
   incompleteness of the manifest image 45ff.
      explanatory incompleteness:
      (WR2) 73
      (WR3) 76
      unstability:
      (WR1) 49
   unreliability of human senses 45 - 46
assumption of separability **165**
belief field **216 - 217**
Bell's inequalities 53ff.
best explanation 171 - 172, 175ff.
best-explaining theory **179 - 180**
   compared to truth 184 - 187
   involved in scientific understanding 183 - 184
better explanatory theory **176 - 177, 179**
better scientific explanatory answer **177, 179**
CF, cognitive field **215**
complete scientific explanatory answer **174**
conceptual order 138
corrective explanation **187 - 188**
de **facto** postulation of unobservable entities by science 45 - 46
degree of truthlikeness **196**
distance measure, d(S,S') **190**
E, environment **153**

epistemic coherence, k(S,S') **190**
epistemic utilities, expected **192**
Ersatz-reference 99 - 100, 116 - 117
explanatory power **181**
explanatory value, EV **178**
f_U, action type function **152**
global explanatory value (GEV) **178**
growth of knowledge 168ff.
incompleteness of the manifest image 45ff.
   see: arguments for scientific realism
inductive systematization 77ff.
intentional action **144**
intentional social action **148**
interpretation of Kantian intuition 102 - 103
J_A, counterpart jointness-function, coupling operator **160**, **165**
j_A, jointness-function **158**
joint social action **145** - **146**
K, composition **153**
K, conduct plan **144**
k(S,S'), epistemic coherence **190**
limit science 12
magic 220ff.
manifest image 10ff., 47ff., 73ff.
   see: incompleteness of the manifest image
mind-independence of objects 108ff.
modifications of Kant's system 38 - 40
Myth of the Given 6, 22ff., 47ff.
   ontological, MG_o **22**, 26 - 27
   epistemic, MG_e **23**, 27 - 28
   linguistic, MG_l **25**, 28 - 31
   compared to metaphysical realism 105 - 106
order of being 13, 129 - 130
order of conceiving 13, 129 - 130
ought-to-be -rules **34-35**, 120 - 121
ought-to-do -rules **34-35**, 120 - 121
parapsychology 230 - 233
pattern governed behavior 33 - 34, 119ff.
P-understandability 177
picturing 115ff.
picturing truth, $truth_p$ **118**
practical inference, schema **171**
prescience 219

protoscience **219 - 220**
pseudoscience 226ff.
R, set of relations defining structure **153 - 154**
rationality in science 218
realism
   arguments for (**see**: arguments for scientific realism)
   internal (**see**: scientific realism)
   metaphysical **96**
   realism and quantum mechanics 53ff.
   scientific (**see**: scientific realism)
reasoning schema (**PR_i**) **150**
religion 223 - 226
   personal god **224**
research field **219**
science
   characterization **216 - 217**
   features of scientific method 211 - 215
scientia mensura 13, 124 - 125, 128 - 129, 132ff., 210 - 211
scientific explanation **172 - 173**
scientific image 10ff., 42ff., 125
scientific realism 4ff.
   causal internal **106 - 107**, 107ff.
   extreme (**ESR**) **126 - 127**
   internal **96**
   minimal (**MSR**) **126**
   moderate (**MOSR**) **126**
   as related to growth of knowledge 200ff.
   Sellarsian (**SSR**) **127**
   strong 50
   weak 49
scientific research activity 211 - 212
scientific understanding **183 - 184**
semantic assertability; truth$_a$ **118**
semantic rules 33
social action 141ff.
stereoscopic (or synoptic) view of the world 14ff.
system σ_B **153 - 154**
systems theory 151ff.
T_A, transition function, growth function **156**, 160, 200
T_{i+1}, successor theory 170ff.
theoretical concepts, role of 87ff.
theoretician's dilemma **65 - 66**

translation function, tr 187 - 188
truth
    asymptotic 169, 194
    correspondence 95 - 96
    epistemic 101ff., 169, **191, 194**
    factual 169
    maximally informative 185 - 186
    truth$_a$ **118 - 119,** 169
    truth$_p$ **118**
unreliability of human   senses 46
use theories of meaning 30ff.
we-intention **147**
X_A, set of states of agent A 156

EPISTEME

A SERIES IN THE FOUNDATIONAL, METHODOLOGICAL,
PHILOSOPHICAL, PSYCHOLOGICAL, SOCIOLOGICAL,
AND POLITICAL ASPECTS OF THE SCIENCES, PURE AND APPLIED

Editor: MARIO BUNGE

Foundations and Philosophy of Science Unit, McGill University

1. William E. Hartnett (ed.), *Foundations of Coding Theory.* 1974, xiii + 216 pp. ISBN 90-277-0536-4.
2. J. Michael Dunn and George Epstein (eds.), *Modern Uses of Multiple-Valued Logic.* 1977, vi + 338 pp. ISBN 90-277-0474-2.
3. William E. Hartnett (ed.), *Systems: Approaches, Theories, Applications.* Including the Proceedings of the Eighth George Hudson Symposium, held at Plattsburgh, New York, April 11–12, 1975. 1977, xiv + 202 pp. ISBN 90-277-0822-3.
4. Wladyslaw Krajewski, *Correspondence Principle and Growth of Science.* 1977, xiv + 138 pp. ISBN 90-277-0770-7.
5. José Leite Lopes and Michel Paty (eds.), *Quantum Mechanics, A Half Century Later.* Papers of a Colloquium on Fifty Years of Quantum Mechanics, held at the University Louis Pasteur, Strasbourg, May 2–4, 1974, x + 310 pp. ISBN 90-277-0784-7.
6. Henry Margenau, *Physics and Philosophy: Selected Essays.* 1978, xxxviii + 404 pp. ISBN 90-277-09001-7.
7. Robert Torretti, *Philosophy of Geometry from Riemann to Poincaré.* 1978, xiv + 459 pp. ISBN 90-277-0920-3.
8. Michael Ruse, *Sociobiology: Sense or Nonsense?* 1979, Second Edition 1985, xvi + 260 pp. ISBN 90-277-1797-4.
9. Mario Bunge, *Scientific Materialism.* 1981, xiv + 224 pp. ISBN 90-277-1304-9.
10. Sal Restivo, *The Social Relations of Physics, Mysticism, and Mathematics.* 1983, viii + 292 pp. ISBN 90-277-1536-X.
11. Joseph Agassi, *Technology.* 1985, xix + 270 pp. ISBN 90-277-2044-4
12. Raimo Tuomela, *Science, Action, and Reality.* 1985, vii + 274 pp. ISBN 90-277-2098-3.